Babak Haghighat

On Topological String Theory with Calabi-Yau Backgrounds

Babak Haghighat

On Topological String Theory with Calabi-Yau Backgrounds

Modularity and Boundary Conditions

Südwestdeutscher Verlag für Hochschulschriften

Impressum / Imprint
Bibliografische Information der Deutschen Nationalbibliothek: Die Deutsche Nationalbibliothek verzeichnet diese Publikation in der Deutschen Nationalbibliografie; detaillierte bibliografische Daten sind im Internet über http://dnb.d-nb.de abrufbar.
Alle in diesem Buch genannten Marken und Produktnamen unterliegen warenzeichen-, marken- oder patentrechtlichem Schutz bzw. sind Warenzeichen oder eingetragene Warenzeichen der jeweiligen Inhaber. Die Wiedergabe von Marken, Produktnamen, Gebrauchsnamen, Handelsnamen, Warenbezeichnungen u.s.w. in diesem Werk berechtigt auch ohne besondere Kennzeichnung nicht zu der Annahme, dass solche Namen im Sinne der Warenzeichen- und Markenschutzgesetzgebung als frei zu betrachten wären und daher von jedermann benutzt werden dürften.

Bibliographic information published by the Deutsche Nationalbibliothek: The Deutsche Nationalbibliothek lists this publication in the Deutsche Nationalbibliografie; detailed bibliographic data are available in the Internet at http://dnb.d-nb.de.
Any brand names and product names mentioned in this book are subject to trademark, brand or patent protection and are trademarks or registered trademarks of their respective holders. The use of brand names, product names, common names, trade names, product descriptions etc. even without a particular marking in this work is in no way to be construed to mean that such names may be regarded as unrestricted in respect of trademark and brand protection legislation and could thus be used by anyone.

Verlag / Publisher:
Südwestdeutscher Verlag für Hochschulschriften
ist ein Imprint der / is a trademark of
OmniScriptum GmbH & Co. KG
Heinrich-Böcking-Str. 6-8, 66121 Saarbrücken, Deutschland / Germany
Email: info@svh-verlag.de

Herstellung: siehe letzte Seite /
Printed at: see last page
ISBN: 978-3-8381-1606-8

Zugl. / Approved by: Bonn, Rheinische Friedrich-Wilhelms-Universitaet, Diss., 2009

Copyright © 2010 OmniScriptum GmbH & Co. KG
Alle Rechte vorbehalten. / All rights reserved. Saarbrücken 2010

Contents

1 Introduction   **5**
  1.1 Motivation 5
  1.2 Outline 8

2 The Physics of the Topological String   **11**
  2.1 Compactifications to $N = 2$ supergravity 11
    2.1.1 Type IIA / IIB string theory in ten dimensions 11
    2.1.2 Compactifications to four dimensions 12
    2.1.3 BPS states 15
  2.2 Gauge theories from geometry 16
    2.2.1 The Seiberg-Witten model 16
    2.2.2 Gauge theories from local Calabi-Yau manifolds 19
  2.3 Counting the entropy of Black Holes 20
    2.3.1 Black Holes in four and five dimensions 21
    2.3.2 Microscopic interpretation of the entropy 22

3 The Topological String   **25**
  3.1 The background geometry 25
    3.1.1 Calabi-Yau manifolds 25
    3.1.2 The Moduli Space 28
  3.2 Supersymmetric nonlinear Sigma Models 36
    3.2.1 $N = (2, 2)$ nonlinear Sigma Models 36
    3.2.2 Linear Sigma Model view point 42
  3.3 Twisting the $N = (2, 2)$ theories 46
    3.3.1 Generalities about topological field theories 46
    3.3.2 The A- and B-twists 49
    3.3.3 Physical Observables of the topological theories 51
    3.3.4 Metric (in)dependence and topological string theory 52
    3.3.5 Dependence on the parameters 54
    3.3.6 The tt^* equations 55
  3.4 The topological A-model 57
    3.4.1 A model without worldsheet gravity 57
    3.4.2 Coupling to topological gravity 61

		3.4.3	Target space perspective	62
		3.4.4	Interpretation around the Conifold singularity	67
	3.5	The topological B-model		68
		3.5.1	B-model without worldsheet gravity	68
		3.5.2	Picard-Fuchs equations	70
		3.5.3	Coupling to topological gravity	72
		3.5.4	The holomorphic anomaly equations	73
	3.6	Mirror Symmetry		75
		3.6.1	Implications for the Topological String	76
	3.7	Solving the holomorphic anomaly equations		78
		3.7.1	The holomorphic limit	78
		3.7.2	Direct Integration	78
		3.7.3	The holomorphic ambiguity	83

4 Grassmannian Calabi-Yau backgrounds — **87**

	4.1	Calabi-Yau complete intersections in Grassmannians		87
		4.1.1	Topological invariants of the manifolds	87
		4.1.2	Plücker embedding	89
		4.1.3	Mirror Construction	90
	4.2	Picard-Fuchs equations for one-parameter models		91
	4.3	The Grassmannian Calabi-Yau $(\mathbb{G}(2,5)\|1,1,3)^1_{-150}$		93
		4.3.1	Picard-Fuchs differential equation and the structure of the moduli space	94
		4.3.2	$g=0$ and $g=1$ Gopakumar-Vafa invariants	95
		4.3.3	Higher genus free energies	95
	4.4	Other Models		99
		4.4.1	$(\mathbb{G}(2,5)\|1,2,2)^1_{-120}$	99
		4.4.2	$(\mathbb{G}(3,6)\|1^6)^1_{-96}$	100
		4.4.3	$(\mathbb{G}(2,6)\|1,1,1,1,2)^1_{-116}$	102
		4.4.4	$(\mathbb{G}(2,7)\|1^7)^1_{-98}$	103
	4.5	5d black hole entropy		103
	4.6	Summary of the models		104

5 Local Calabi-Yau backgrounds — **107**

	5.1	Local Mirror Symmetry		107
		5.1.1	The local A-model	108
		5.1.2	The local B-model	109
	5.2	Direct Integration in local Calabi-Yau geometries		110
	5.3	$\mathbb{K}_{\mathbb{P}^1 \times \mathbb{P}^1} = \mathcal{O}(-2,-2) \to \mathbb{P}^1 \times \mathbb{P}^1$		111
		5.3.1	Review of the moduli space $\mathcal{M}$	113
		5.3.2	Solving the topological string on local $\mathbb{F}_0$ at large radius	113
		5.3.3	Solving the topological string on local $\mathbb{F}_0$ at the conifold locus	117
		5.3.4	Solving the topological string on local $\mathbb{F}_0$ at the orbifold point	119

5.3.5 Relation to the family of elliptic curves 120
5.4 $\mathbb{K}_{\mathbb{F}_1} = \mathcal{O}(-2,-3) \to \mathbb{F}_1$ 121
    5.4.1 Solving the topological string on local $\mathbb{F}_1$ at large radius 122
    5.4.2 Solving the topological string on local $\mathbb{F}_1$ at the conifold locus . . . 124
    5.4.3 Relation to the family of elliptic curves 125

6 K3 fibrations   127
6.1 Calabi-Yau hypersurfaces in toric varieties 127
6.2 Picard-Fuchs equations and the B-model 129
6.3 Moduli Space of K3 Fibrations 131
    6.3.1 K3 Fibrations 131
    6.3.2 The Moduli Space of the Mirror 132
6.4 Physical boundary conditions 134
    6.4.1 The Strong coupling singularity 134
    6.4.2 The weak coupling divisor and meromorphic modular forms 136
    6.4.3 The Seiberg-Witten plane 141
    6.4.4 The Gepner point 142
6.5 Solution of the Models 143
    6.5.1 $M_1 = \mathbb{P}_4^{(1,1,2,2,2)}[8]$ 143

7 Conclusions   151

A Yukawa-couplings from Picard-Fuchs operators   155

B Modular anomaly versus holomorphic anomaly   157
B.1 $PSL(2,\mathbb{Z})$ modular forms 158

C Details: Grassmannian Calabi-Yau manifolds   161
C.1 Chern classes and topological invariants 162
C.2 Tables of Gopakumar-Vafa invariants 163
C.3 5D black hole asymptotic 169

D Details: Local Calabi-Yau manifolds   171
D.1 Gopakumar-Vafa invariants of local Calabi-Yau manifolds 172

E Details: K3 Fibrations   175
E.1 Gopakumar-Vafa invariants 176

Chapter 1

Introduction

1.1 Motivation

One of the major problems of modern theoretical physics, both conceptually as well as practically, is the reconcilliation of the two pillars of 20th century physics, namely gravitation and quantum mechanics. The theory of gravitation is described in an astonishingly elegant way by Einstein's theory of general relativity and has thus far passed a few but impressive experimental tests. These tests are mainly of astrophysical and cosmological nature, the most prominent of which being the manifestation of the dynamical nature of space-time through Hubble's observation of an expanding universe. The conceptual core of general relativity in the context of the dialectic development of physical theories is the unification of Newton's theory of gravitation with special relativity, being itself a synthesis of Maxwell's electromagnetism and the concept of inertial systems. On the other hand the history of quantum mechanics has taken a different road on the landscape of physical theories, having it's origin in atomic physics as well as the physics of radiation and light. This time it was not a contradiction between physical theories which led to the birth of the new theory but rather a contradiction between classical theories compared to experimental observations about the radiation of black bodies and the spectrum of light emitted by individual atoms. Quantum mechanics found its climax in the development of quantum field theory which is capable of explaining the interactions between individual particles to an accuracy unprecedented among the predictive power of human theories.

Today we are standing in front of the conceptual incompatibility between quantum field theory and general relativity. As a dynamical classical theory general relativity admits a Lagrangian formulation and its dynamical variables expanded around a classical solution can be interpreted as quantum fields giving rise to a spin two particle known as the graviton. However, it turns out that this quantum field theory is ill defined in the sense that it is not renormalizable. Here, we should point out that within the so called asymptotic safety program, a lot of efforts are being devoted to establishing the existence of an ultraviolet fixed point at which Quantum Einstein Gravity can be renormalized. Up to now the results of these efforts are still speculative. Nonrenormalizability of general relativity makes practical calculations concerning the quantum nature of space-time at the time right after the big bang and near black holes impossible from the beginning. Furthermore, it is known from the work of Hawking and Bekenstein, that a black hole is a thermodynamical object with an entropy which goes linearly with the area of its horizon. However, there is no way to look behind the horizon of a black hole, even theoretically, to find out what the microscopic states are which give rise to the macroscopic entropy and radiation. These issues can only be addressed in a theory of quantum gravity where the interactions of gravitons with the microscopic objects forming the black hole are well described. Switching the point of view to the one of contemporary particle physics we find similar difficulties in explaining the observed richness of particle spectra and interactions. One of the major problems of the Standard Model of particle physics are the large quantum corrections contributing to the square of the Higgs mass, known as the hierarchy problem. One way to solve it is to introduce a new symmetry, namely supersymmetry, which doubles the particle spectrum and thus provides a mechanism to cancel the quadratic divergencies appearing in loop corrections to the Higgs mass. There are also other problems like the explanation of the amount of dark matter in the universe as well as gauge coupling unification at the GUT scale, for which solutions can be found within the supersymmetric extension of the standard model.

String theory is a modern theoretical approach which incorporates both, quantum gravity as well as supersymmetry. It regularizes field theory by introducing a new scale, known as the *string scale*, and it naturally incorporates gravitation. Furthermore, it is a supersymmetric theory providing the framework for constructing supersymmetric models. In string theory the one-dimensional trajectory of a particle in spacetime is replaced by a two-dimensional orbit of a string denoted by *worldsheet*. In mathematical terms the worldsheet is a Riemann surface, i.e. a complex manifold with complex dimension one. The situation is very similar to the quantum mechanical case where the introduction of Planck's constant $\hbar$ is responsible for passing from classical to quantum physics. In a similar manner, in string theory one introduces a new fundamental constant $\alpha' \sim (10^{-32}\text{cm})^2$ being a parameter for the tension of the string. It turns out that this way gravity is regularized as it is no longer possible to probe spacetime beneath distances of order $\sqrt{\alpha'} \sim 10^{-32}\text{cm}$. In other words there is an absolute minimum uncertainty in length. Classical field theory results and in particular general relativity arise then in the limit $\alpha' \to 0$. However, it turns out that string theory is only consistent, i.e. anomaly free, in a spacetime with ten dimensions. This makes it unavoidable to compactify the theory on a six dimensional compact space to establish contact with our observed four dimensional world. In order to preserve the amount of supersymmetry relevant for phenomenology the compactification manifold has to satisfy several conditions, namely it must be complex and Kähler and allow for one covariantly constant spinor. Although being very stringent, these conditions lead to a vast number of solutions known as Calabi-Yau manifolds. Each such space leads to a different four dimensional particle content and interactions and thus to a different physical vacuum state. It is conjectured that all these vacua are connected through the notion of extremal transitions [1]. The major challenge of string phenomenology is then to find the right vacuum describing our universe at the current state of evolution. On the other hand string theory also suffers from some drawbacks

1.1. MOTIVATION

from the conceptual point of view. It is known that there is not only one string theoretic construction but there are five consistent theories at once, all being known only in their perturbative regimes. But since 1995 it has become clear that all five theories are connected through a chain of dualities and to an eleven-dimensional theory called M theory which is only known in its low energy limit as eleven dimensional supergravity. This picture involves the existence of new extended objects known as Dp-branes which are p-dimensional analogs of the string but different in some important properties. The best understood duality is the so called mirror symmetry which relates type IIA string theory compactified on a Calabi-Yau manifold M to type IIB string theory on a "mirror" Calabi-Yau W. In the extremely simplified case where M is a circle of radius R, W would be a circle of radius α'/R. The symmetry states that the two theories compactified in such a way admit the same particle spectrum and four dimensional physics. We will make extensive use of mirror symmetry in this thesis. The other symmetry we will be needing for our calculations is the so called heterotic - type II duality which is a symmetry of a completely different nature. Here the string coupling constant g_s in the one theory which is a field theoretical quantity is related to the size of a sphere in the other theory which is a purely geometric quantity. To test such dualities BPS states become very important as these are the only states in the theory which prevail and are protected against corrections even in the nonperturbative regime.

One of the major breakthroughs of string theory in recent years is the calculation of the Bekenstein-Hawking entropy for supersymmetric black holes in terms of a string theoretic microscopic description. Such an entropy calculation is possible for black holes which are extremal in the sense that their charge and their mass are in a fixed relation. On the string theory side this involves a counting of BPS states which are realized as D-branes wrapping certain cycles of the compactification manifold.

Another important line of research in string theory is the construction of four dimensional supersymmetric vacua by choosing a certain compactification geometry. This involves calculating the four dimensional effective action together with its superpotential and prepotential.

There are several tools available in string theory to address the above questions, each of which emphasizing a different viewpoint. One of these tools is the topological string. It is a simplified version of the critical string theory in which the path integral localizes on the topological subsector of the theory. Physically this is a simplified approximation. Within the sigma model representation of the critical string the path integral is an integral over all possible maps of the two dimensional worldsheet to the target space which in general is a complex three dimensional Calabi-Yau manifold tensored with Minkowski spacetime. Whereas in the topological string the space of integration is reduced to the space of the distinct classical solutions. One may ask why one should look at such a simplified version of the original theory. For this question there are two main answers. First of all in its standard formulation string theories exhibit an infinite number of fields and particles, an infinite dimensional symmetry algebra which is only very vaguely understood and where a lot of these symmetries are "broken". There is a claim that the topological version is another phase of the physical theory, in which much more symmetries are preserved and unbroken and in which the spectrum is considerably simpler. Therefore one hopes that from the examination of the one theory one gets clues about some principals of the physical theory. The other important advantage of the topological string is that it is able to compute certain physical amplitudes of the real string theory. These are terms of the effective four-dimensional theory which depend holomorphically on the moduli. The most important examples of such terms are the superpotential, the prepotential and the gauge kinetic function of $N = 1$ and $N = 2$ supersymmetric field theories. Furthermore, topological string theory counts naturally certain BPS states and is therefore ideally suited for computing the microscopic entropy of extremal spinning five dimensional black holes. Within the OSV conjecture [2] there are clues that the theory is also of great relevance for the entropy of four dimensional supersymmetric black holes.

There are two main approaches present for solving the topological string on Calabi-Yau backgrounds. The first approach is called the topological A model and has also great relevance for the mathematical point of view. In the A model the classical solutions around which the string path integral localizes are holomorphic maps from the Riemann surface of the world sheet to curves of the Calabi-Yau target space. These are instantons labeled by the genus g of the holomorphically embedded curve and the degrees d_i which count the number of intersection with the divisors of the Calabi-Yau. The mathematical tool to calculate the number of such maps is called localization. It makes use of the fact that the *ambient space*, being the space in which the Calabi-Yau manifold is embedded, admits a group action (i.e. $(\mathbb{C}^*)^n$ in case of a n dimensional toric variety) in order to localize the path integral to fix points of the group action. The disadvantage of the A model is that it only provides solutions at the large volume point in moduli space. The second approach is called the topological B model and rests heavily on the use of mirror symmetry. Here one performs all calculations in the mirror Calabi-Yau knowing that there the classical solutions are just maps from the worldsheet to points of the target space which are much easier to control. Then one "translates" the result by mirror symmetry to the original Calabi-Yau where they can be interpreted as A model results. The calculation on the B model side makes use of the properties of the topological free energies around boundary divisors of the mirror Calabi-Yau moduli space where physical descriptions of the particle spectrum are available. Such information can consist of the number and type of particles becoming massless at the relevant divisor as well as phase transitions going through an enhanced symmetry point. In this thesis we shall follow the second approach to topological string theory, i.e. the B model, as this provides us with solutions on the whole of moduli space.

1.2 Outline

This thesis is organized as follows.

In chapter 2 we review the physics lying behind and the physics captured by topological string theory. That is we start with an introduction to type II supergravity where we compactify the ten dimensional bosonic actions down to four dimensions. This way it is possible to identify the four dimensional multiplets with the moduli fields of the com-

1.2. OUTLINE

pactification manifold. Next, we pass over to the description of gauge theories within the framework of $N = 2$ supersymmetry where in particular we concentrate on the Seiberg-Witten solution of the $SU(2)$ gauge theory. Having reviewed this construction we present its embedding into string theory and comment on how the topological string captures certain nonperturbative aspects of these gauge theories. Finally we discuss four and five dimensional supersymmetric black holes, their macroscopic entropy, and their embedding into string theory together with a microscopic interpretation of the entropy.

Chapter 3 is devoted to an introduction of the main ideas and calculational tools behind the topological string. This implies a presentation of complex geometry and the notion of Calabi-Yau manifolds as relevant target spaces. Here we will include a section about the moduli space of Calabi-Yau manifolds and special geometry as this is of great relevance for later discussions. Then we will pass over to present a review of supersymmetric sigma models. Here we describe $N = (2,2)$ world sheet supersymmetry compactifications and some details of the $N = (2,2)$ CFT. Furthermore, we include a section about the linear sigma model perspective which represents a unifying framework for various geometric constructions and phases in string theory. Following these discussions there will be a description of the A and B twists leading to topological theories. Finally the topological models are combined with a Calabi-Yau target space and coupled to topological gravity. Here we subdivide again between the A and the B model. Having discussed the topological A and B models we turn our attention to mirror symmetry, where we will explain the mirror map and the genus zero sector. Last but not least, we will pass over to solving the holomorphic anomaly equations. The first step is to find a recursive solution of the equations genus by genus and the second will be to explain the boundary conditions at various points in moduli space.

In chapter 4, solutions of the topological string on Calabi-Yau manifolds which are complete intersections in Grassmannians are presented. These results are also published in [3]. The chapter starts with an introduction to Grassmannian varieties and the notion of Calabi-Yau complete intersections in these. Next, mirror symmetry for these spaces is reviewed. Then we pass on to present results of the solutions to the anomaly equations expanded on boundary points of the moduli space. Having derived topological amplitudes for this class of spaces up to genus 5 we turn our attention to the black hole interpretation and analyze the discrepancy between the microscopic and macroscopic entropy evaluations.

The next chapter, chapter 5, analyzes the solutions to the anomaly equations on local Calabi-Yau manifolds. The results of this part were published in [4]. First we give a toric description for this class of manifolds and identify a Riemann surface as their mirror. Then the direct integration procedure for these spaces is described and the boundary information from conifold expansions is extracted. The last section clarifies the relations between topological amplitudes and in particular generators of the amplitudes with modular forms.

Chapter 6 deals with the topological string on K3 fibrations. The results are published in [5]. The first section deals with the description of compact toric varieties and the construction of Calabi-Yau hypersurfaces in them. An important application of the toric description, which we shall review, is the derivation of Picard-Fuchs equations from symmetries of the ambient space. The second section is concerned with the duality of the Heterotic String with the type II string and the importance of K3 fibrations in this context. Then, we give an overview over the moduli space and identify all relevant boundary divisors. Having described the geometry of the moduli space we summarize the physical boundary conditions at the various divisors and present expansions of the free energies. The last section deals with the question of integrability on these spaces.

The last chapter contains some concluding remarks and directions for the future. The appendices A, B contain supplementary material about the calculation of Yukawa-couplings and definitions for $SL(2, \mathbb{Z})$ modular forms. The last three appendices, namely C, D, E, contain the tables of BPS degeneracies, a plot about the macroscopic and microscopic black hole entropies for Grassmannians and a table for the topological invariants of Grassmannian Calabi-Yau three-folds.

Chapter 2

The Physics of the Topological String

This chapter is meant to be an overview about the physical principles underlying the topological string and its main physical applications. To this respect we first review the compactification of type IIA/IIB string theory on Calabi-Yau manifolds. Then we pass over to the discussion of $N = 2$ gauge theories and how one can obtain them from the choice of the compactification geometry. Here we present the Seiberg-Witten solution and its relevance for topological string computations. Last but not least we turn to a short exposition of macroscopic and microscopic black hole physics.

2.1 Compactifications to $N = 2$ supergravity

2.1.1 Type IIA / IIB string theory in ten dimensions

Superstring theories are only consistent as quantum theories and anomaly free in ten spacetime dimensions. Their perturbative description is given by a supersymmetric sigma model with target space a ten dimensional manifold which usually splits up into $\mathbb{R}^{1,3} \times M$, where M is denoted by *compactification space*. In the limit of large volume of M and small string coupling the interacting theory captured by the sigma model reduces to supergravity, i.e. a quantum field theory. The nonperturbative sector of string theory contains solitonic states, namely the D-branes, which also admit a supergravity description in terms of p-forms. The nature of the nonperturbative description of string theory is still not clear and therefore a full description of D-branes and their bound states remains far away. However, in many phenomenological applications it is convenient to consider first the supergravity limit of string theory and include nonperturbative corrections as a second step. This said, we want to describe in the following the supergravity picture arising from string theory and its impact on four dimensional physics.

We shall focus our exposition on the bosonic sector of low energy effective actions of type IIA and type IIB string theory. That is, we will be dealing with type IIA and type IIB supergravity in 10 dimensions. Both theories are $N = 2$ supersymmetric, the difference being that in the one theory the gravitino multiplet has opposite chirality to the gravitino sitting in the graviton multiplet[1].

Starting with the non-chiral type IIA theory, its massless spectrum comprises the metric $\hat{g}_{MN}$, the two-from $\hat{B}_2$, the dilaton $\hat{\phi}$, and a one and a three-form denoted by $\hat{A}_1$ and $\hat{C}_3$. Note that the fermionic components follow by supersymmetry. The bosonic action of this theory is given by [7]

$$S_{IIA} = \int \left[e^{-2\hat{\phi}} \left(-\frac{1}{2}\hat{R} * \mathbf{1} + 2d\hat{\phi} \wedge *d\hat{\phi} - \frac{1}{4}\hat{H}_3 \wedge *\hat{H}_3 \right) \right.$$
$$\left. - \frac{1}{2} \left(\hat{F}_2 \wedge *\hat{F}_2 + \hat{F}_4 \wedge *\hat{F}_4 \right) + \mathcal{L}_{top} \right], \quad (2.1.1)$$

where the fields strengths are defined as

$$\hat{F}_2 = d\hat{A}_1, \quad \hat{F}_4 = d\hat{C}_3 - \hat{B}_2 \wedge d\hat{A}_1, \quad \hat{H}_3 = d\hat{B}_2, \quad (2.1.2)$$

and we will ignore the topological terms $\mathcal{L}_{top}$ as they are not of particular importance for the argumentation we want to carry out.

The type IIB theory is the chiral type II theory and its massless bosonic fields are given by the metric $\hat{g}_{MN}$, the two-form $\hat{B}_2$, the dilaton $\hat{\phi}$, and the zero, two and four-forms l, $\hat{C}_2$ and $\hat{A}_4$. Speaking in string theoretic terms, one sees that the two theories differ only in their RR sectors while their NS-NS sectors comprising the fields $\hat{g}_{MN}$, $\hat{B}_2$ and $\hat{\phi}$ are equal. As the RR sector contains only even forms in type IIB the action will only contain odd form field strengths. Its bosonic part is

$$S_{IIB} = \int e^{-2\hat{\phi}} \left(-\frac{1}{2}\hat{R} * \mathbf{1} + 2\hat{\phi} \wedge *d\hat{\phi} - \frac{1}{4}\hat{H}_3 \wedge *\hat{H}_3 \right)$$
$$- \frac{1}{2} \left(dl \wedge *dl + \hat{F}_3 \wedge *\hat{F}_3 + \frac{1}{2}\hat{F}_5 \wedge *\hat{F}_5 \right) - \frac{1}{2}\hat{A}_4 \wedge \hat{H}_3 \wedge d\hat{C}_2, \quad (2.1.3)$$

where the field strengths are defined as

$$\hat{H}_3 = d\hat{B}_2, \quad \hat{F}_3 = d\hat{C}_2 - ld\hat{B}_2, \quad \hat{F}_5 = d\hat{A}_4 - \frac{1}{2}\hat{C}_2 \wedge d\hat{B}_2 + \frac{1}{2}\hat{B}_2 \wedge d\hat{C}_2. \quad (2.1.4)$$

2.1.2 Compactifications to four dimensions

To obtain a four dimensional physical theory the ten dimensional supergravities have to be compactified on a manifold M. One can choose the amount of preserved supersymmetry by the choice of the holonomy of the internal manifold M. The number of left supersymmetries will be equal to the number of spinors which can be chosen to be singlets under the holonomy group. A spinor is an irreducible representation of the algebra $\mathfrak{so}(1, d-1)$ and has dimension $2^{d/2}$ for d even and $2^{(d-1)/2}$ for d odd. Furthermore, a spinor may be

---
[1]The latter is known as the *chiral* theory and the former as the *non-chiral*

2.1. COMPACTIFICATIONS TO $N = 2$ SUPERGRAVITY

real ($\mathbb{R}$), complex ($\mathbb{C}$), or quaternionic ($\mathbb{H}$) depending on d, see [8] for more details. The general rule is

$$\begin{array}{ll} \mathbb{R} & \text{if} \quad d = 1, 2, 3 (\mathrm{mod}\ 8) \\ \mathbb{C} & \text{if} \quad d = 0 (\mathrm{mod}\ 4) \\ \mathbb{H} & \text{if} \quad d = 5, 6, 7 (\mathrm{mod}\ 8). \end{array}$$

A complex representation has twice as many degrees of freedom as a real representation and a quaternionic representation has the same number of degrees of freedom as a complex representation due to constraints. To see what happens when one compactifies type II supergravity down to a lower number of dimensions replace the space $\mathbb{R}^{1,d_0-1}$ by $\mathbb{R}^{1,d_1-1} \times M$, for M a compact space of dimension $d_0 - d_1$. Then one has to consider how a spinor of $\mathfrak{so}(1, d_0 - 1)$ decomposes under the maximal subalgebra $\mathfrak{so}(1, d_1 - 1) \oplus \mathfrak{so}(d_0 - d_1) \subset \mathfrak{so}(1, d_0-1)$. The holonomy of M acts on this maximal subalgebra and each representation which is invariant under its action will lead a new supersymmetry in the compactified target space. For $N = 2$ in ten dimensions the general rule is the following

Holonomy of M	N in $D = 4$
$SO(6)$	8
$SU(2)$	4
$\mathbb{Z}_2 \times SU(2)$, $SU(3)$	2

Table 2.1.1: Holonomy and supersymmetry.

Here N is the number of supersymmetries and D the dimension of spacetime. An example for the first type is the torus T^6, for the second case one can choose the target space to be $K3 \times T^2$, and the last case comprises any *Calabi-Yau* manifold. A mathematical definition of K3 surfaces and Calabi-Yau manifolds will be given in section 3.1. As our main interest lies in theories with $N = 2$ supersymmetry we will look at type II compactifications on Calabi-Yau manifolds in more detail.

Compactification on Calabi-Yau manifolds

The field content of $N = 2$ supersymmetry in four dimensions can be constructed from the multiplets of $N = 1$ supersymmetry. An $N = 2$ hypermultiplet is build from two chiral multiplets resulting in two complex scalars and two fermions. On the other hand a vector and a chiral superfield together give rise to an $N = 2$ vector multiplet, its field content being a vector, two gaugini and one complex scalar. The $N = 2$ graviton multiplet is the union of the $N = 1$ graviton and gravitino multiplets. We summarize these observations in table 2.1.2.

Let us start by compactifying type IIA on a Calabi-Yau. We want to identify the bosonic field content of the resulting theory with the bosonic fields presented in table 2.1.2. In a reduction of a higher dimensional theory to a lower dimensional one, denoted by the term *Kaluza-Klein*-compactification, one expands the higher dimensional fields in terms of harmonics of the compact space in order to only keep massless modes in the

Multiplet	Bosons	Fermions
hyper-multiplet	$4 \times \phi$	$2 \times \psi$
vector	A_μ, Φ	$2 \times \lambda$
graviton	$g_{\mu\nu}, A^0_\mu$	$2 \times \Psi_\mu$

Table 2.1.2: $N=2$ supermultiplets in four dimensions. The symbol ϕ is reserved for real scalars and Φ for complex ones.

effective theory. Expanding in this spirit the ten dimensional fields $\hat{A}_1$, $\hat{B}_2$ and $\hat{C}_3$ in Calabi-Yau harmonic forms one obtains [2]

$$\begin{aligned}
\hat{A}_1 &= A^0, \\
\hat{B}_2 &= B_2 + b^i \omega_i, \\
\hat{C}_3 &= C_3 + A^i \wedge \omega_i + \xi^A \alpha_A - \tilde{\xi}_A \beta^A,
\end{aligned} \quad (2.1.5)$$

where C_3 is a three-form, B_2 a two-form, (A^0, A^i) are one-forms and b^i, ξ^A, $\tilde{\xi}_A$ are scalar fields in $D=4$. Note that $\omega^i, i=1,\cdots,h^{1,1}(M)$ are harmonic 2-forms and α_A, β^A, $A=0,\cdots,h^{2,1}$, are harmonic three-forms of the internal manifold M. However, these are not yet all fields appearing in the four dimensional theory. There are also massless modes associated to metric deformations of the internal geometry. In the case of Calabi-Yau manifolds these are Kähler and complex structure deformations denoted by $v^i, i=1,\cdots,h^{1,1}$ and $z^a, a=1,\cdots,h^{2,1}$, as will be described in more detail in section 3.1.2. The b^i and v^i combine together to form the complex fields $t^i = b^i + iv^i$. These scalars together with the one-forms A^i form the bosonic content of $h^{1,1}$ vector multiplets. Turning our attention to the complex structure deformations we see that the complex fields z^a and the scalars $\xi^a, \tilde{\xi}_a$ form together exactly the bosonic content of an $N=2$ hypermultiplet, namely 4 scalar bosons. The remaining fields adjust themselves into the tensor and the gravitational multiplet.

Next, we pass over to the compactification of type IIB theory. Again, in order to derive the massless spectrum the ten dimensional fields are expanded into Calabi-Yau harmonic forms

$$\begin{aligned}
\hat{B}_2 &= B_2 + b^i \omega_i, \quad \hat{C}_2 = C_2 + c^i \omega_i, \\
\hat{A}_4 &= D_2^i \wedge \omega_i + V^A \alpha_A - U_B \beta^B + \rho_i \tilde{\omega}^i,
\end{aligned} \quad (2.1.6)$$

where now in addition to the harmonics already present in the type IIA case, this time we also have harmonic $(2,2)$-forms $\tilde{\omega}_i, i=1,\cdots,h^{1,1}$ being dual to the ω_i introduced earlier. Here one has to note that the bosonic fields D_2^i and ρ_i are duals of one another and the one-form fields (V^A, U_A) are related by electric-magnetic duality. Thus we only have to consider half of these fields. We choose the scalar fields $\rho_i, 1=1,\cdots,h^{1,1}$, and the vector fields $V^A, A=0,\cdots,h^{2,1}$, to be physical. Combining these with the Calabi-Yau moduli we see that one is left with $h^{1,1}$ hyper-multiplets (ρ^i, v^i, b^i, c^i) and $h^{2,1}$ vector multiplets (V^a, z^a).

---

[2] Such a reduction was performed the first time in [9]

2.1. COMPACTIFICATIONS TO $N = 2$ SUPERGRAVITY

Comparing with the type IIA spectrum in four dimensions we see that the number of hyper- and vector multiplets are exchanged. This observation is the supergravity origin of mirror symmetry. In the full string theory picture these two moduli spaces receive quantum corrections and thus make mirror symmetry a far more nontrivial statement. Looking at the vector moduli spaces, the type IIA side gets quantum corrected by *worldsheet instantons* while the type IIB vector moduli space remains uncorrected. Worldsheet instantons can be interpreted as *BPS states* arising from D-brane bound states which in turn have important applications to nonperturbative aspects of string theory. Indeed, in this thesis we will make extensive use of mirror symmetry to calculate the degeneracy of specific BPS states and hence we will use the next subsection to introduce them briefly to the reader.

2.1.3 BPS states

BPS states are massive supersymmetric states which play an important role in the understanding of the nonperturbative nature of Superstring theory as their properties are protected against corrections even in the nonperturbative regime. Let us present their definition and their main properties in a short exposition here, for more detail we refer to the lecture notes [10]. Consider a theory with N supersymmetries where $N = 2r$ for some r. Diagonalizing the anti-symmetric central charge matrix $Z^{ab} = -Z^{ba}$ of the theory into blocks of 2×2, we obtain

$$Z = \mathrm{diag}(\epsilon Z_1, \cdots, \epsilon Z_r) \quad \epsilon^{12} = -\epsilon^{21} = 1, \qquad (2.1.7)$$

where the $Z_{\bar{a}}, \bar{a} = 1, \cdots, r$ are called the *real central charges*. Next, we look at the massive representations of the theory and define creation and annihilation operators by $\mathcal{Q}_{\alpha\pm}^{\bar{a}} \equiv \frac{1}{2}(Q_{\alpha}^{1\bar{a}} \pm \sigma_{\alpha\dot{\beta}}^{0}(Q_{\dot{\beta}}^{2\bar{a}})^{\dagger})$ and their hermitian conjugates. The only nontrivial supersymmetry algebra relation left is then

$$\left\{\mathcal{Q}_{\alpha\pm}^{\bar{a}}, (\mathcal{Q}_{\beta\pm}^{\bar{b}})^{\dagger}\right\} = \delta_{\bar{b}}^{\bar{a}} \delta_{\alpha}^{\beta}(M \pm Z_{\bar{a}}). \qquad (2.1.8)$$

The left hand side (2.1.8) must be positive for any unitary representation of the supersymmetry algebra. This immediately gives us the so called *BPS bound* for the mass of the particles in the spectrum

$$M \geq |Z_{\bar{a}}| \quad \bar{a} = 1, \cdots, r = [N/2]. \qquad (2.1.9)$$

For configurations with $|Z_{\bar{a}}| = M$ the BPS bound is saturated and one of the supercharges $\mathcal{Q}_{\alpha+}^{\bar{a}}$ or $\mathcal{Q}_{\alpha-}^{\bar{a}}$ must vanish. As a consequence we obtain a shorter supersymmetry representation, i.e. the phenomenon of *multiplet shortening* occurs. If $M = |Z_{\bar{a}}|$ for $\bar{a} = 1, \cdots, r_0$, and $M > |Z_{\bar{a}}|$ for other values of $\bar{a}$, the corresponding supersymmetry representation has dimension 2^{2N-2r_0} and is denoted by $1/2^{r_0}$ BPS. For $N = 2$ supersymmetry in four dimensions we list the number of irreducible spin representation for BPS saturated multiplets in table 2.1.3.

CHAPTER 2. THE PHYSICS OF THE TOPOLOGICAL STRING

spin ≤ 1	0	1/2	1
$N = 2$ BPS hyper	2	1	0
$N = 2$ BPS vector	1	2	1

Table 2.1.3: Number of irreducible representations as a function of spin

The names hyper and vector arise from state counting which shows that they have the same number of states as massless hyper- and massless vector multiplets.

The condition $M = |Z_{\bar{a}}|$ will remain valid even in the strong coupling regime and will not suffer corrections as one does not expect short multiplets to turn into the full multiplets, with many more states!

2.2 Gauge theories from geometry

Having described four dimensional supergravity we turn next to supersymmetric gauge theories. We will see that in string theory there exists a mechanism to decouple gravity from these theories and thus obtain a purely gauge theoretic description. In the following we will first describe the Seiberg-Witten gauge theory as it gives rise to many interesting features, and as a second step we will explain its embedding into string theory.

2.2.1 The Seiberg-Witten model

The setup

Consider $N = 2$ supersymmetric Yang-Mills theory in four dimensions. Assume that the gauge theory is $SU(2)$ with one vector supermultiplet $\mathcal{A}$. Then the particle content of $\mathcal{A}$ is, according to section (2.1.2), given by an $N = 1$ chiral multiplet, whose components we shall denote by ϕ and ψ, and an $N = 1$ vector multiplet with components λ and one gauge field A_μ. All fields come in the adjoint representation. Under the global $SU(2)_R$ symmetry the bosonic fields A_μ and ϕ are singlets and λ, ψ form a doublet. Furthermore, there is an additional $U(1)_R$ symmetry acting on the fields ϕ,ψ. However, quantum mechanically this R-symmetry is broken to its $\mathbb{Z}_8$ subgroup by an anomaly in the theory we are considering. In $N = 1$ superspace formalism, the Lagrangian is expressed as

$$\frac{1}{4\pi}\text{Im}\left[\int d^4\theta \frac{\partial \mathcal{F}(A)}{\partial A}\overline{A} + \int d^2\theta \frac{1}{2}\frac{\partial^2 \mathcal{F}(A)}{\partial A^2}W_\alpha W^\alpha\right], \quad (2.2.1)$$

where A is the $N = 1$ chiral multiplet in the $N = 2$ vector multiplet $\mathcal{A}$, and its scalar component we denote by a. The prepotential $\mathcal{F}$ gives rise to the Kähler potential

$$K = \text{Im}\left(\frac{\partial \mathcal{F}(A)}{\partial A}\overline{A}\right). \quad (2.2.2)$$

2.2. GAUGE THEORIES FROM GEOMETRY

Note that we have not included a superpotential, but the D-term gives rise to the following classical scalar potential

$$V(\phi) = \frac{1}{g^2}\text{Tr}[\phi,\phi^\dagger]^2. \tag{2.2.3}$$

So, classically the vacua of the theory are given by the configurations where ϕ and $\phi^\dagger$ commute. In our case the gauge group is $SU(2)$ and thus we can take $\phi = \frac{1}{2}a\sigma^3$, with $\sigma^3 = \text{diag}(1,-1)$ and a a complex parameter. As $SU(2)$ acts by its Weyl group on the field a, sending it to $-a$, we see that the gauge inequivalent vacua are parametrized by the gauge invariant quantity $u = \frac{1}{2}a^2 = \text{Tr}\phi^2$. For non-zero a supersymmetry remains unbroken while the gauge symmetry is broken to $U(1)$ and the global $\mathbb{Z}_8$ symmetry is broken to $\mathbb{Z}_4$.

The solution

Seiberg and Witten [11] have presented a solution to the effective infrared limit of the theory, described by the Wilsonian action. That is, they have constructed the space of effective quantum corrected gauge inequivalent vacua and have deduced the particle spectrum from it. Their Ansatz relies on three major considerations

- holomorphy,
- global symmetries,
- the existence of a nonsingular weak coupling limit.

As all quantities of interest can be deduced from the *holomorphic* prepotential $\mathcal{F}$, the goal will be to compute its full quantum corrected expansion. The perturbative corrections to $\mathcal{F}$ were already deduced in [12]. The tree level and one loop contributions add up to

$$\mathcal{F}_{oneloop} = i\frac{1}{2\pi}\mathcal{A}^2\ln\frac{\mathcal{A}^2}{\Lambda}, \tag{2.2.4}$$

where Λ is the dynamically generated scale and all higher loop contributions vanish. The logarithm in the expression is responsible for the anomalous transformation behavior of the $U(1)_R$. Further corrections arise from instantons. The new terms have to be invariant under the remaining $\mathbb{Z}_4$ R-symmetry which suggests the holomorphic Ansatz

$$\mathcal{F} = i\frac{1}{2\pi}\mathcal{A}^2\ln\frac{\mathcal{A}^2}{\Lambda^2} + \sum_{k=1}^{\infty}\mathcal{F}_k\left(\frac{\Lambda}{\mathcal{A}}\right)^{4k}\mathcal{A}^2, \tag{2.2.5}$$

where the k'th term arises as a contribution of k instantons. Negative powers of k are absent as they would violate the existence of a nonsingular weak coupling limit. It will turn out that infinitely many of the $\mathcal{F}_k$ are nonzero. Seiberg and Witten deduce the instanton corrected prepotential by looking at the metric on the space of vacua (the moduli space) and computing its behavior around specific points in moduli space. The metric is given by

$$(ds)^2 = \text{Im}\tau(a) da d\bar{a}, \qquad (2.2.6)$$

where $\tau(a)$ is a holomorphic function $\tau = \partial^2 \mathcal{F}/\partial a^2$. The first major observation is that Imτ cannot be a globally defined smooth function as it would cease to be positive definite. As such it must have singularities on the moduli space leading to *monodromies* around singular points. A correct description of the metric uses the variables $a_D = \partial \mathcal{F}/\partial a$ and a, in terms of which we have

$$(ds)^2 = \text{Im} da_D d\bar{a} = -\frac{i}{2}(da_D d\bar{a} - da d\bar{a}_D). \qquad (2.2.7)$$

If we parameterize the moduli space by a variable u (corresponding to Trϕ^2), the functions a and a_D will transform nontrivially by going around singular points on the u-plane. Indeed one can show that these *monodromy transformations* form a subgroup of $SL(2, \mathbb{Z})$ denoted by Γ_2. What is the origin of these monodromies and how can one compute them? The answer lies at the heart of the physics. Seiberg and Witten conjecture that the singularities come from massive particles of spin $\leq 1/2$ that become massless at particular points in the moduli space. Moreover, these particles are not elementary but bound states and correspond to monopoles and dyons. These are BPS states whose mass is given by the formula $M = \sqrt{2}|n_m a_D + n_e a|$. Let us look at a monopole with mass a_D which is becoming massless at the point $a_D(u_0) = 0$. Then the Wilsonian effective action will incorporate the effect of integrating out such a massless particle. More precisely, using the one loop beta function one can show that the magnetic coupling is

$$\tau_D \sim -\frac{i}{\pi}\ln a_D. \qquad (2.2.8)$$

Using this result and the relation between the functions a and a_D described in [11] one can show that by going in a loop around the point u_0 in the u-plane a and a_D transform as

$$a_D \to a_D \qquad (2.2.9)$$
$$a \to a - 2a_D. \qquad (2.2.10)$$

A similar behavior shows up around two other points in moduli space: the massless dyon point and the weak coupling point $u = \infty$.

The picture unraveled here is the one of a *Riemann surface* with periods a and a_D! Using the symmetries of the theory and the details of the monodromy behavior just described this Riemann surface can be identified to be described by the equation

$$y^2 = (x - 1)(x + 1)(x - u). \qquad (2.2.11)$$

The singular points $x = 1$ and $x = -1$ correspond the massless monopole and the massless dyon point. This beautiful picture finally solves the initial problem as deducing the prepotential associated to the moduli space of this Riemann surface is equivalent to computing the instanton corrected holomorphic function $\mathcal{F}$ of the $N = 2$ theory.

2.2. GAUGE THEORIES FROM GEOMETRY

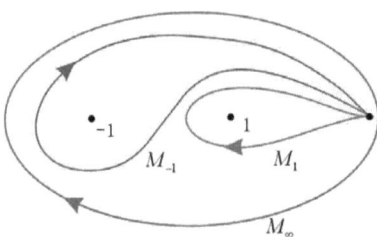

Figure 2.1: The Seiberg-Witten u-plane with a choice of base point.

2.2.2 Gauge theories from local Calabi-Yau manifolds

The Seiberg-Witten $SU(2)$ gauge theory can be embedded into string theory in the sense that it can be obtained in a certain limit from a Calabi-Yau compactification. This was analyzed in [13]. The main idea can be traced back to the observation that in type IIA compactifications over *K3 fibrations* (a K3 surface fibred over $\mathbb{P}^1 = S^2$) *ADE singularities* of K3 lead to enhanced gauge symmetry of ADE type [14]. The reason is that 2-branes of type IIA wrapped around vanishing 2-cycles lead to precisely the missing states expected for gauge symmetry enhancement.

In the case of the $SU(2)$ theory the geometry of the Calabi-Yau will consist of a base $\mathbb{P}^1$ with fibre in the singular limit being $\mathbb{C}^2/\mathbb{Z}_2$. Blowing up $\mathbb{C}^2/\mathbb{Z}_2$ we see that the geometry locally contains a fibration of $\mathbb{P}^1$ over $\mathbb{P}^1$. The $W^\pm$ will correspond to 2-branes wrapped around the $\mathbb{P}^1$ fibre and their mass is proportional to the area of the 2-sphere. Furthermore, $1/g^2$ is proportional to the area of the base sphere, where g is coupling constant of the gauge theory. Now, the nature of the particular limit taken in [13] is sending $M_{planck} \to \infty$ in order to decouple gravity and obtain a pure gauge theory. Geometrically this is realized by sending size of the base, denoted by t_b, to infinity, i.e. $t_b \to \infty$, and the size of the $\mathbb{P}^1$ fibre, denoted by t_f, to zero. However, as we have

$$\exp(-1/g^2) = \exp(-t_b) \sim \epsilon^4 \Lambda^4$$
$$t_f \sim \epsilon a, \qquad (2.2.12)$$

we have to ensure that the ratio $\frac{\exp(-1/g)}{t_f^4}$ stays finite in order to obtain the finite instanton contributions in (2.2.5) while sending $\epsilon \to 0$.

Of what use is the topological string here? The genus 0 topological free energy, denoted by F^0, encodes all instanton corrections of the type IIA vector moduli space. Due to mirror symmetry it can be computed classically on the mirror manifold and then translated back to the type IIA side. As it turns out, in the local limit we are employing here the mirror manifold is characterized by a Riemann surface which turns out to be exactly the same as the Seiberg-Witten curve (2.2.11). Therefore, the genus 0 topological string free energy contains all instanton contributions of the gauge theory. This result can be used to obtain systematically nonperturbative corrections to quantum field theories.

A further generalization of the setup just presented, apart from going to higher gauge groups, is to couple the gauge theory to matter. This is done by including hypermultiplets in the adjoint or in the fundamental representation. Geometrically the inclusion of r hypermultiplets in the adjoint is *engineered* by fibering ADE singularities over a complex curve of genus r (see figure 2.2).

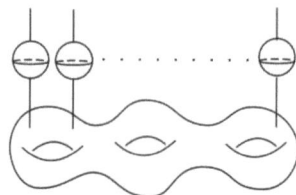

Figure 2.2: Illustration of an $N = 2$ $SU(2)$ gauge theory with 3 hypermultiplets in the adjoint.

2.3 Counting the entropy of Black Holes

Now we shall leave the path of pure gauge theories and return to the supergravity picture. This is the correct arena for analyzing the theory of black holes. From the point of view of the topological string this means including the higher free energies $F^g(t_i)$ as these correspond to the following F-terms in the effective four dimensional $N = 2$ supergravity

$$\int d^4 x F^g(t_i) R_+^2 F_+^{2g-2}. \tag{2.3.1}$$

Here, R_+ is the self-dual part of the Riemann tensor and F_+ is the self-dual part of the graviphoton field strength. The couplings $F^g(t_i)$ depend on the vector moduli arising from compactification of type IIA string theory on a Calabi-Yau manifold. The general rule to compute the F^g is very similar to the one used in the Seiberg-Witten solution. The terms (2.3.1) arise in the Wilsonian effective action by integrating out massive states. However, on certain points on the moduli space of the Calabi-Yau manifold some of the states integrated out become massless and lead to singularities in the effective four dimensional theory. In the case of the topological string these are BPS states corresponding to D-brane bound states. A knowledge of the expansion of the $F^g(t_i)$ around singular points in the moduli space and of the monodromy of the periods there can be restrictive enough to fix the F^g completely. This has important applications for the theory of black holes to which we shall turn next.

2.3.1 Black Holes in four and five dimensions

One of the central goals of string theory is to provide a quantum version of general relativity. An immediate application and consistency check of such a theory would be the counting of the microscopic entropy of black holes, i.e. providing a quantum explanation for the Bekenstein-Hawking entropy formula. Thus far, this has been achieved for so called *extremal* black holes, i.e. charged black holes with their mass equaling their charge. One of the major breakthroughs was the work of Strominger and Vafa [15]. Let us start by reviewing the classical geometry of the Reissner-Nördstrom (RN) black hole. It is a time independent, spherically symmetric solution of Einstein gravity coupled to the electromagnetic field. The solution of the metric and vector field is as follows:

$$ds^2 = -\left(1 - \frac{2G_N M}{r} + \frac{G_N Q^2}{r^2}\right) dt^2 + \left(1 - \frac{2G_N M}{r} + \frac{G_N Q^2}{r^2}\right)^{-1} dr^2 + r^2 d\Omega^2,$$

$$A_0 = \frac{Q}{r}, \quad A_i = 0, i = 1, 2, 3, \qquad (2.3.2)$$

where A_μ are the spacetime components of the vector potential and Q denotes the charge of the solution. There are two coordinate singularities ($g_{rr} = \infty$) at $r = r_+$ (outer horizon) and $r = r_-$ (inner horizon)

$$r_\pm = G_N M \pm \sqrt{(G_N M)^2 - G_N Q^2}, \qquad (2.3.3)$$

where the event horizon is given by the outer horizon $r = r_+$. When $M = |Q|/\sqrt{G_N}$, r_+ coincides with r_- and the black hole is called extremal. Such a solution embedded into supergravity is called BPS as the charges correspond to the central charges of the supersymmetry algebra. One can see that such a black hole is completely characterized by its charge. When looking at spinning black holes with angular momentum J the situation is slightly different. Here, supersymmetric configurations, i.e. configurations which preserve half of the supersymmetry, are achieved for $M = |Q|/\sqrt{G_N}$ for arbitrary value of angular momentum J. However, the extreme limit for a spinning black hole is reached at $M^2 - |Q|^2/G_N = J^2$ and therefore the solution does not have any unbroken supersymmetry. On the other hand, lifting the situation to $5D$, rotating extremal black hole solutions have been constructed which do have unbroken supersymmetries. The metric of a rotating black hole with one half of the supersymmetries of $N = 2$ supergravity in five dimensions is given by

$$ds^2 = (1 - \frac{\mu}{r^2})^2 \left[dt - \frac{4J \sin^2\theta}{\pi(r^2 - \mu)} d\phi + \frac{4J \cos^2\theta}{\pi(r^2 - \mu)} d\psi \right]^2$$
$$- (1 - \frac{\mu}{r^2})^{-2} dr^2 - r^2(d\theta^2 + \sin^2\theta d\phi^2 + \cos^2\theta d\psi^2). \qquad (2.3.4)$$

One sees that for the rotating solutions near the horizon at $r^2 \to \mu \equiv r_0^2$ the metric does not split into a product space as there are non-diagonal components. Furthermore, the metric for the three-sphere is distorted and one has

$$r_0^2 d^2\Omega_3(J) = r_0^2 \left(d^2\Omega_3 - \left(\frac{4J}{r_0^3 \pi}\right)^2 (\sin^2\theta d\phi - \cos^2\theta d\psi)^2 \right). \qquad (2.3.5)$$

The volume of the distorted 3-sphere defines the area of the horizon of the rotating black hole

$$A(J) = 2\pi^2 \sqrt{r_0^6 - J^2}. \qquad (2.3.6)$$

In order to deduce the Bekenstein-Hawking entropy $S_{BH} = A/(4G_N)$ one first has to fix the value of r_0. To this respect one property of such black hole configurations is very useful, namely extremal supersymmetric black holes behave as attractors [16, 17]. That is, the moduli take fixed values at the horizon which depend only on the charges and not on the values of the moduli at infinity. Furthermore, the area of these black holes can be found by extremizing the value of the central charge in moduli space. Using this attractor mechanism the authors of [17] derive the following relation

$$S_0 = 2\pi \sqrt{Q^3 - J^2}, \qquad (2.3.7)$$

where Q is the graviphoton charge of the black hole. We have given the entropy the index 0 to stress the fact that this result is only valid in classical regime. This result will receive further corrections from higher derivative interactions according to Wald's formula [18].

2.3.2 Microscopic interpretation of the entropy

In order to present a microscopic interpretation for black hole entropy, we first have to embed the supergravity picture into string theory. In fact, in the case of 5 dimensional rotating extremal black holes, the corresponding supergravity solution is obtained by compactifying M theory on a Calabi-Yau threefold X. Then the black hole will be characterized by a charge $Q \in H_2(X, \mathbb{Z})$ and $SU(2)_L \subset SO(4)$ angular momentum J. Microscopically, a $5d$ black hole with membrane charge $Q \in H_2(X, \mathbb{Z})$ is engineered by wrapping $M2$ branes around the two-cycle Q. The result of this in $5d$ is a supersymmetric spectrum of BPS states which are labeled by Q and by their spin content (j_L, j_R). One further has to sum over j_R with an insertion of $(-1)^{2j_R}$ which finally gives as resulting spectrum for a membrane charge Q

$$R_Q = \sum_{r=0}^{g} n_Q^r I_{r+1}, \qquad (2.3.8)$$

where

$$I_l = \left[2(\mathbf{0}) + \left(\frac{1}{2}\right)\right]^l \qquad (2.3.9)$$

encodes the spin content j_L, and n_Q^r are called Gopakumar-Vafa invariants. The n_Q^r are computed by the topological string and can be extracted by the knowledge of the free energies of the genera $0 \leq g \leq r$. For a given charge Q there are only finitely many nonzero n_Q^r. As argued in [19] one can write down the following generating function for the supersymmetric degeneracies of BPS states with membrane charge Q

$$\sum_{J,Q} \Omega(Q, J) = \sum_Q \text{tr}_{R_Q}(-1)^{2j_L} y^{j_L}. \qquad (2.3.10)$$

2.3. COUNTING THE ENTROPY OF BLACK HOLES

This way the $\Omega(Q, J)$ can be extracted to be

$$\Omega(Q, J) = \sum_r \binom{2r+2}{J+r+1} n_Q^r, \qquad (2.3.11)$$

where $J = 2j_L$. As the black hole entropy is given by the logarithm of the number of microstates we arrive at

$$S(Q, J) = \log(\Omega(Q, J)). \qquad (2.3.12)$$

This should agree with the macroscopic result in the large charge limit $Q \gg 1$ and $Q \gg J$.

Now we have talked enough about the applications of the topological string. It is time to explain how to actually compute topological amplitudes in string theory. This will be the purpose of the rest of this thesis.

Chapter 3
The Topological String

This chapter forms the theoretical grounding on which the calculations of later chapters rest. As a first step we will introduce the background geometry relevant for us, namely Calabi-Yau manifolds and their moduli spaces. Then we come to the description of nonlinear sigma models on such spaces which are part of the perturbative description of string theory. Twisting the nonlinear sigma model gives rise to topological field theories, namely the A and B model. The succeeding sections deal with a detailed description of these two models and their coupling to gravity. In the case of the B model the coupling to gravity leads to the holomorphic anomaly equations for which a solution method is presented in the last section.

3.1 The background geometry

3.1.1 Calabi-Yau manifolds

The background geometries of the topological string in the critical case consist of so called *Calabi-Yau* manifolds of complex dimension 3 which we shall discuss in this section. As we will move on the ground of complex geometry we refer the reader to the excellent treatises [21, 22] for mathematical details. Let us start with a first definition which says that a Calabi-Yau manifold X is a complex, Kähler manifold with vanishing first chern class $c_1(\mathcal{T}_X) = 0$. On a complex manifold the expansion of the total Chern class of the tangent bundle $\mathcal{T}_X$ reads

$$c(\mathcal{T}_X) = 1 + \sum_j c_j(\mathcal{T}_X) = \det(1 + \mathcal{R}) = 1 + \operatorname{tr}\mathcal{R} + \operatorname{tr}(\mathcal{R} \wedge \mathcal{R} - 2(\operatorname{tr}\mathcal{R})^2) + \cdots. \quad (3.1.1)$$

Here $\mathcal{R}$ is the curvature two form which in complex coordinates is given by

$$\mathcal{R} = iR^k_{l i \bar{j}} dz^i \wedge d\bar{z}^{\bar{j}}, \quad (3.1.2)$$

Written in this form the curvature two-form can be thought of as the curvature of the holomorphic tangent bundle $\mathcal{T}_X = TX^{(1,0)}$ endowed with the spin connection. In this picture the indices k and l arise from the Lie Algebra matrix which acts on vectors of the bundle. Taking the trace over the above form leads to the so called *Ricci form* which is a closed form as can be checked easily. The first chern class is a cohomology element defined by the linear term in the expansion (3.1.1), namely $c_1(X) = [\text{tr}\mathcal{R}/2\pi] \in H^2(X,\mathbb{R})$. Therefore, we see that a vanishing Ricci form also implies a vanishing first Chern class. However, the converse is a very nontrivial theorem proved by Yau which says that if the first Chern class vanishes then X admits a Ricci-flat metric. The exact form of the statement is:

Theorem(Calabi-Yau). If X is a complex Kähler manifold with vanishing first Chern class and with Kähler form ω, then there exists a unique Ricci-flat metric on X whose Kähler form ω' is in the same cohomology class as ω.

Let us now present some equivalent definitions of a Calabi-Yau manifold which will prove useful in different circumstances. The first is concerned with the question of holonomy. Note that a complex Kähler manifold always has a holonomy group which is contained in $U(n)$, n being the complex dimension of the manifold. This is due to the closeness property of the Kähler form ω which translates to the covariant constancy of the almost complex structure J which in turn requires the holonomy group to commute with J. In order to further constrain the holonomy to $SU(n)$ the $U(1)$ part of the spin connection must be set to zero. But the $U(1)$ part of the spin connection is exactly the Ricci form. Therefore we see that a Kähler manifold with vanishing first Chern class has holonomy $SU(n)$. Utilizing the *holonomy principle* this leads us to a further equivalent definition of a Calabi-Yau manifold. The assumption that the holonomy group is contained in $SU(n)$ says that any chosen trivialization of $\Lambda^n T_p^* X^{(1,0)} = \det(T_p^* X^{(1,0)})$ is left invariant by parallel transport. Hence there exists a covariantly constant section Ω of the canonical bundle K_X. The holomorphic three form Ω is without zeros as the zero section itself is covariantly constant. Hence, K_X is trivialized by Ω. Similarly one shows that there are no parallel sections of the kth power of the cotangent bundle where $1 < k < n$. This implies that $h^{0,k} = h^{k,0} = 0$. In summary we obtain the following form for the Hodge diamond of Calabi-Yau manifolds

$$\begin{array}{ccccccc} & & & 1 & & & \\ & & 0 & & 0 & & \\ & 0 & & h^{1,1} & & 0 & \\ 1 & & h^{2,1} & & h^{2,1} & & 1 \\ & 0 & & h^{1,1} & & 0 & \\ & & 0 & & 0 & & \\ & & & 1 & & & \end{array}, \qquad (3.1.3)$$

where we have restricted to the three dimensional case. There is a specialty in the case of complex surfaces, i.e. Calabi-Yau manifolds which are of complex dimension two. Here one can use the Hirzebruch-Riemann-Roch theorem to compute

$$\chi(X,\mathcal{O}_X) = \int_X \text{ch}(\mathcal{O}_X)\text{td}(X) = \int_X \text{td}(X) = \int_X \frac{c_1^2(X) + c_2(X)}{12}. \qquad (3.1.4)$$

3.1. THE BACKGROUND GEOMETRY

On the other hand we have $\chi(X, \mathcal{O}_X) = h^0(\mathcal{O}_X) - h^1(\mathcal{O}_X) + h^2(\mathcal{O}_X) = h^{0,0}(X) - h^{0,1}(X) + h^{0,2}(X)$ which is equal to 2 from the above discussion. As $c_1(X) = 0$ by definition we obtain

$$2 = \int_X \frac{c_2(X)}{12} = \frac{1}{12}\chi(X), \qquad (3.1.5)$$

thus $\chi(X) = 24$. This puts severe constrains on the topology of such manifolds and one finds that up to diffeomorphism there is only one such Calabi-Yau, called the $K3$ surface. Combining all results, its Hodge diamond is given by

$$\begin{matrix} & & 1 & & \\ & 0 & & 0 & \\ 1 & & 20 & & 1 \\ & 0 & & 0 & \\ & & 1 & & \end{matrix} \quad . \qquad (3.1.6)$$

Let us now turn over to a concrete example. Consider the hypersurface in $\mathbb{C}P^4$ given by the locus $P(x_1, \cdots, x_5) = \sum_i x_i^5 = 0$ where $x_1, \cdots, x_n$ are the homogeneous coordinates of the projective space. This hypersurface is called the Quintic hypersurface as the polynomial P is of degree 5. In order to see why it is Calabi-Yau we have to go one step back and analyze the chern classes of projective space. These are deduced by making use of the *Euler sequence*

$$0 \to \mathbb{C} \to H^{\oplus(n+1)} \to \mathcal{T}_{\mathbb{C}P^n} \to 0, \qquad (3.1.7)$$

to show that

$$c(\mathcal{T}_{\mathbb{C}P^n}) = c(H^{\oplus(n+1)}) = c(H)^{n+1} = (1+\omega)^{n+1}, \qquad (3.1.8)$$

where H is the hyperplane bundle and ω its first Chern class. Specializing to the case of $\mathbb{C}P^4$ we get

$$c(\mathbb{C}P^4) = (1+\omega)^5. \qquad (3.1.9)$$

Here one should note that the right hand side is subject to $\omega^5 = 0$ as the manifold is four-dimensional. A direct consequence of the above formula is that $c_1(\mathbb{C}P^4) = 5\omega$. To proceed further, we have to find a way to relate the Chern class of $X = \{P = 0\}$ in $\mathbb{C}P^4$ to the one of $\mathbb{C}P^4$ itself. Such a relation is obtained by noting that

$$\mathcal{T}_{\mathbb{C}P^4}|_{P=0} = \mathcal{T}_X \oplus \mathcal{N}_{X/\mathbb{C}P^4}, \qquad (3.1.10)$$

where $\mathcal{T}_X$ and $\mathcal{N}_{X/\mathbb{C}P^4}$ are the tangent bundles of X and the normal bundle of X inside of $\mathbb{C}P^4$. Together with the Whitney product formula this gives

$$c(\mathcal{T}_X) = \frac{c(\mathcal{T}_{\mathbb{C}P^4})}{c(\mathcal{N}_{X/\mathbb{C}P^4})}. \qquad (3.1.11)$$

The only ingredient still missing is the Chern class of the normal bundle. It can be shown that it is given by the degree of the polynomial P, i.e. in our case $c(\mathcal{N}_{X/\mathbb{C}P^4}) = (1+5\omega)$. Interpreting the right hand side of (3.1.11) as a formal power series in ω we find

$$c(\mathcal{T}_X) = \frac{(1+\omega)^5}{(1+5\omega)} = 1 + (5-5)\omega + \cdots. \qquad (3.1.12)$$

Thus, we have proved that $c_1(\mathcal{T}_X) = 0$ and a quintic hypersurface in $\mathbb{C}P^4$ is a Calabi-Yau manifold with complex dimension 3. We can carry out the discussion even a bit further. Expanding (3.1.11) to third order, we see that $c_3 = -40\omega^3$ and using $\int_X \omega^3 = \int_{\mathbb{C}P^4} \omega^4 = 5$ we obtain $\chi(X) = \int_X c_3(\mathcal{T}_X) = -200$.

There are numerous examples of Calabi-Yau manifolds and many generalizations of the above construction. In this thesis we will be dealing with models which are described through several constraints in Grassmannian spaces, with local toric Calabi-Yau manifolds and with manifolds which are hypersurfaces in weighted projective space.

3.1.2 The Moduli Space

Yau's theorem suggests to view the parameter space of Calabi-Yau manifolds as the parameter space of Ricci-flat Kähler metrics. In the following we will analyze the consequences of this picture for the metric deformations $\delta g_{\mu\nu}$ where we will follow [24]. Let $g_{\mu\nu}$ be a Ricci-flat metric for X and $g_{\mu\nu} + \delta g_{\mu\nu}$ a perturbation of the former so that the Ricci-flatness is still fulfilled. Then we have

$$R_{\mu\nu}(g) = 0, \quad R_{\mu\nu}(g + \delta g) = 0. \tag{3.1.13}$$

Using the gauge $\nabla^\nu g_{\mu\nu} = 0$ and expanding (3.1.13) to first order in δg one finds that $\delta g_{\mu\nu}$ satisfies the *Lichnerowicz* equation

$$\nabla^\lambda \nabla_\lambda \delta g_{\mu\nu} + 2R_\mu{}^\kappa{}_\nu{}^\tau \delta g_{\kappa\tau} = 0. \tag{3.1.14}$$

The special properties of Kähler manifolds imply that this equation is satisfied by metric perturbations of mixed type $\delta g_{m\bar{n}}$, and of pure type δg_{mn}, $\delta g_{\bar{m}\bar{n}}$, separately. We can associate to the variations of mixed type the $(1,1)$-form

$$i\delta g_{m\bar{n}} dx^m \wedge dx^{\bar{n}}, \tag{3.1.15}$$

which is harmonic if and only if the metric variation satisfies equation (3.1.13). Similarly metric variations of the pure type may be contracted with the unique holomorphic three-form to yield the $(2,1)$-form

$$\Omega_{ij}{}^{\bar{n}} \delta g_{\bar{m}\bar{n}} dx^i \wedge dx^j \wedge dx^{\bar{m}} \tag{3.1.16}$$

which is again harmonic due to (3.1.13). Thus, we have mapped the zero modes of the Lichnerowicz equation to elements of $H^{1,1}(X)$ and $H^{2,1}(X)$. As is already evident the zero modes of mixed type correspond to variations of the Kähler class and give rise to b_{11} parameters. On the other hand variations of pure type correspond to complex structure deformations. This can be seen as follows. As each variation transforms X again into a Kähler manifold there must be a coordinate system in which the pure parts of the metric vanish. But under coordinate transformations $x^m \mapsto x^m + f^m(x)$ we have

$$\delta g_{\bar{m}\bar{n}} \mapsto \delta g_{\bar{m}\bar{n}} - \frac{\partial f^r}{\partial \bar{x}^{\bar{m}}} g_{r\bar{n}} - \frac{\partial f^r}{\partial \bar{x}^{\bar{n}}} g_{\bar{m}r} \tag{3.1.17}$$

3.1. THE BACKGROUND GEOMETRY

For those transformations which keep the complex structure fixed, f is a holomorphic function, and therefore in this case there is no impact on the metric. In other words the pure part of metric variations can only be removed through non holomorphic coordinate transformations which change the complex structure.

Consider the example of the previous section, i.e. the quintic hypersurface in $\mathbb{C}P^4$. As we have not specified the form of the polynomial P each coefficient in front of a monomial in P maps into the space of complex structure deformation parameters. Although this map is surjective it is not injective as many of these coefficients turn out to be equivalent by coordinate redefinitions. These are given by maps $x_i \mapsto M_i^j x_j$ with $M \in GL(5,\mathbb{C})$. Counting the monomials of a degree five polynomial and the generators of $GL(5,\mathbb{C})$ one sees that there are $126 - 25 = 101$ complex structure deformation parameters.

Complex structure moduli space

Let us parameterize the complex structure deformations by parameters z^a, $a = 1, \cdots, h^{2,1}$. Then we can define

$$\chi_{aij\bar{k}} = -\frac{1}{2}\Omega_{ij}{}^{\bar{l}}\frac{\partial g_{\bar{k}\bar{l}}}{\partial z^a}, \quad \chi_a = \frac{1}{2}\chi_{aij\bar{k}}dx^i \wedge dx^\lambda \wedge dx^{\bar{k}}, \qquad (3.1.18)$$

where each $\chi_a \in H^{2,1}(X)$. There is yet another way to see how elements of $H^{2,1}(X)$ parameterize complex structure deformations. The inverse relation is given by

$$\delta g_{\bar{k}\bar{l}} = -\frac{1}{||\Omega||^2}\bar{\Omega}_{\bar{k}}{}^{mn}\chi_{amn\bar{l}}\delta z^a, \qquad (3.1.19)$$

where we have defined $||\Omega||^2 := \frac{1}{3!}\Omega_{lmn}\bar{\Omega}^{lmn}$. Now we are ready to obtain an expression for the metric on the complex structure moduli space. This is done by integrating the square of the metric variations over the whole Calabi-Yau, i.e.

$$\begin{aligned}2G_{a\bar{b}}\delta z^a \delta z^{\bar{b}} &\equiv \frac{1}{2V}\int_X g^{k\bar{l}}g^{m\bar{n}}\delta g_{km}\delta g_{\bar{l}\bar{n}}g^{\frac{1}{2}}d^6x \\ &= -\frac{2i}{V||\Omega||^2}\delta z^a \delta z^{\bar{b}}\int_X \chi_a \wedge \overline{\chi_b},\end{aligned} \qquad (3.1.20)$$

where V is the volume of the manifold. From this we conclude

$$G_{a\bar{b}} = -\frac{\int_X \chi_a \wedge \overline{\chi_b}}{\int_X \Omega \wedge \overline{\Omega}}. \qquad (3.1.21)$$

The metric $G_{a\bar{b}}$ is called the Weil-Peterson metric. It admits certain special properties to which we shall turn in the following. First of all note that we can write Ω in the form $\Omega = \frac{1}{3!}h(f)\epsilon_{ijk}df^i df^j df^k$, where $f^i(x,z)$ are holomorphic coordinates varying with z such that $f^i(x,z_0) = x^i$. Under a change of complex structure $df^i(x,z)$ becomes partly of type $(1,0)$ and partly of type $(0,1)$. Together with the fact that the exterior derivative d commutes with the variation of complex structure $\frac{\partial}{\partial z^a}$ we then obtain the relation

$$\frac{\partial \Omega}{\partial z^a} \in H^{3,0}(X) \oplus H^{2,1}(X). \qquad (3.1.22)$$

It can be shown that the $(2,1)$ part is equal to χ_a, thus one gets

$$\frac{\partial \Omega}{\partial z^a} = k_a \Omega + \chi_a, \qquad (3.1.23)$$

where k^a is only a function of the z^a as the space $H^{3,0}$ is one dimensional. A direct consequence of (3.1.23) is that the Weil-Peterson metric can be derived from a potential, i.e. it can be written in the form

$$G_{a\bar{b}} = -\frac{\partial}{\partial z^a}\frac{\partial}{\partial z^{\bar{b}}} \log(i \int_X \Omega \wedge \overline{\Omega}), \qquad (3.1.24)$$

which shows that the space of complex structures is Kähler with Kähler potential

$$-\log(i \int_X \Omega \wedge \overline{\Omega}). \qquad (3.1.25)$$

Next, we want to derive an expression for (3.1.25) in terms of the periods of the holomorphic 3-form. Let $(A^a, B_b), a, b = 0, \cdots, h^{2,1}$ be a canonical homology basis for $H_3(X, \mathbb{Z})$ and (α_a, β^b) be the dual cohomology basis defined by

$$\int_{A^a} \alpha_a = \int_X \alpha_a \wedge \beta^b = \delta_a^b, \quad \int_X \beta^b = \int_X \beta^b \wedge \alpha_a = -\delta_a^b. \qquad (3.1.26)$$

In terms of this basis the following integrals define the periods of Ω

$$X^a \equiv \int_{A^a} \Omega, \quad \mathcal{F}_a \equiv \int_{B_a} \Omega. \qquad (3.1.27)$$

The X^a are homogeneous projective coordinates of the complex structure moduli space and Ω can be viewed as being homogeneous of degree 1 in these coordinates [25]. This implies $\mathcal{F}_a = \mathcal{F}_a(X)$. Looking at the above definitions we find that

$$\Omega = X^a \alpha_a - \mathcal{F}_a(X) \beta^a, \qquad (3.1.28)$$

which establishes an expansion of the holomorphic 3-form in terms of its periods. Together with the identity

$$\int_X (\Omega \wedge \frac{\partial \Omega}{\partial X^a}) = 0, \qquad (3.1.29)$$

which is a direct consequence of (3.1.23), this yields

$$2\mathcal{F}_a = \frac{\partial}{\partial X^a}(X^c \mathcal{F}_c). \qquad (3.1.30)$$

We see that $\mathcal{F}_a$ is the gradient of a function which is homogeneous of degree two

$$\mathcal{F}_a = \frac{\partial \mathcal{F}}{\partial X^a}, \quad \mathcal{F}(\lambda X) = \lambda^2 \mathcal{F}(X). \qquad (3.1.31)$$

3.1. THE BACKGROUND GEOMETRY

In the light of these results we can return to the formula for the Kähler potential (3.1.25) and express it in terms of the periods in the following way

$$e^{-K} = -i\left(X^a \frac{\partial \overline{\mathcal{F}}}{\partial \overline{X}^a} - \overline{X}^a \frac{\partial \mathcal{F}}{\partial X^a}\right), \tag{3.1.32}$$

which establishes the manifold as of *special Kähler type* with $\mathcal{F}$ being the *prepotential*. All identities which we shall establish in the following as a consequence of (3.1.32) are properties of the so called *special geometry*. Defining affine coordinates

$$t^i = \frac{X^i}{X^0}, \quad i = 1, \cdots, h^{2,1}, \tag{3.1.33}$$

we can rewrite the Kähler potential as

$$-\log\left(i\int_X \Omega \wedge \overline{\Omega}\right) = -\log\left(i\left[2(\mathcal{F} - \overline{\mathcal{F}}) - (t^i - \bar{t}^i)\left(\frac{\mathcal{F}}{\partial t^i} + \frac{\partial \overline{\mathcal{F}}}{\partial \bar{t}^i}\right)\right]\right). \tag{3.1.34}$$

Let us now state some formulae which will become important in the following sections. First of all note that taking derivatives of the holomorphic three form Ω with respect to the X^a yields

$$\frac{\partial}{\partial X^a}\Omega \in H^{3,0} \oplus H^{2,1}$$

$$\frac{\partial^2}{\partial X^a \partial X^b}\Omega \in H^{3,0} \oplus H^{2,1} \oplus H^{1,2} \tag{3.1.35}$$

$$\frac{\partial^3}{\partial X^a \partial X^b \partial X^c}\Omega \in H^{3,0} \oplus H^{2,1} \oplus H^{1,2} \oplus H^{0,3},$$

where we have used again the fact that each complex structure derivative transforms a $(1,0)$-form into a linear combination of $(1,0)$- and $(0,1)$-forms and vice versa. For example the last summand in the last line of (3.1.35) is produced only if each of the three derivatives with respect to the complex structure parameters $\frac{\partial}{\partial X^a}$ hits one df^i in $\Omega = \frac{1}{3!}h(f)\epsilon_{ijk}df^i df^k df^l$. Equations (3.1.35) can be rewritten in a simple way once we introduce the spaces

$$F^p = \bigoplus_{i \geq p} H^{i,k-i}, \quad k = \dim_{\mathbb{C}} X = 3. \tag{3.1.36}$$

With this notation the right hand side of the first line in (3.1.35) is equal to F^2, the right hand side of the second line is F^1 and the right hand side of the third line corresponds to F^0. Furthermore, note that the subspace $H^{3,0}(X)$ within $H^3(X, \mathbb{C})$ defines a line bundle over the complex structure moduli space which we shall denote by $\mathcal{L}$ from now on. A choice of the holomorphic three-form Ω defines a section of this line bundle. For a section of $\mathcal{L}$, the action of the gauge transformation (Kähler transformation) is parametrized by a holomorphic function $f(t)$ and expressed as

$$K(t, \bar{t}) \to K(t, \bar{t}) - \log f(t) - \log \bar{f}(\bar{t}), \quad \Omega \to f(t)\Omega. \tag{3.1.37}$$

Therefore, we see that the covariant derivative of a section $h(t)$ of the line bundle $\mathcal{L}^n$ is defined as
$$D_t h = \partial_t + n(\partial_t K)h. \quad (3.1.38)$$
A consequence of the nonzero $(0,3)$-part of $\frac{\partial^3}{\partial X^a \partial X^b \partial X^c}\Omega$ is that we can define the so called *Yukawa coupling*

$$\begin{aligned} C_{ijk} &= \int_X \Omega \wedge \frac{\partial^3}{\partial X^i \partial X^j \partial X^k}\Omega \\ &= \frac{\partial^3 \mathcal{F}}{\partial X^i \partial X^j \partial X^k} \\ &= (X^0)^2 \frac{\partial^3}{\partial t^i \partial t^j \partial t^k}\mathcal{F}(t), \quad i,j,k \in \{1,\cdots,h^{2,1}\} \end{aligned} \quad (3.1.39)$$

This is a *pseudo*-topological invariant in that it does depend on the complex structure of X. The first line in (3.1.39) transforms covariantly with respect to coordinate transformations $X^i \mapsto z^i(X)$. This is due to the fact that we get only nonzero contributions from terms where all derivatives with respect to z act on $\Omega(z)$ and none acts on terms of the form $\frac{\partial z^i}{\partial X^j}$. Therefore, we can rewrite the triple couplings in terms of periods which provide expressions valid in every coordinate system

$$C_{ijk} = \int_X \Omega \wedge \partial_i \partial_j \partial_k \Omega = \sum_{a=0}^{h^{2,1}} (X^a \partial_i \partial_j \partial_k \mathcal{F}_a - \mathcal{F}_a \partial_i \partial_j \partial_k X^a). \quad (3.1.40)$$

One can see from the first equality in equation 3.1.40 that the Yukawa coupling is a section of $\text{Sym}^3(T\mathcal{M}) \otimes \mathcal{L}^2$ where by $\mathcal{M}$ we denote the complex structure moduli space. Another important consequence of special geometry is the following relation connecting the antiholomorphic derivative of the metric connection to the three-point couplings

$$\bar{\partial}_{\bar{\imath}} \Gamma^k_{ij} = \delta^k_i G_{j\bar{\imath}} + \delta^k_j G_{i\bar{\imath}} - C_{ijl} \bar{C}^{kl}_{\bar{\imath}}, \quad (3.1.41)$$

where we have $\bar{C}^{kl}_{\bar{\imath}} = e^{2K} G^{k\bar{k}} G^{l\bar{l}} C_{\bar{k}\bar{l}\bar{\imath}}$.

Up to here we have focused on the local properties of the complex structure moduli space. Let us now briefly comment on the global form of such a space. For ease of explanation consider the case where the Calabi-Yau is given by the vanishing locus of a homogeneous polynomial P of degree d in weighted projective four-dimensional space $\mathbb{P}_4^{(k_1,k_2,k_3,k_4,k_5)}$. In this case the Calabi-Yau condition translates into the identity $d = \sum_i k_i$. The most general form for P would then be

$$P = \sum_{i_1,i_2,i_3,i_4,i_5} a_{i_1 i_2 i_3 i_4 i_5} x_1^{i_1} x_2^{i_2} x_3^{i_3} x_4^{i_4} x_5^{i_5}, \quad (3.1.42)$$

where $\sum_j k_j i_j = d$ and the $(x_1,\cdots,x_5)$ are homogeneous coordinates of the weighted projective space. Different choices for the constants $a_{i_1 i_2 \cdots i_5}$ correspond to different choices for the complex structure of the underlying Calabi-Yau manifold. However, as in the case

3.1. THE BACKGROUND GEOMETRY

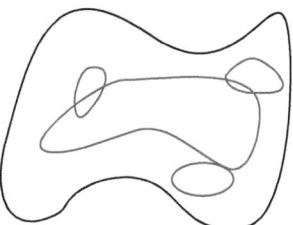

Figure 3.1: The complex structure moduli space

of the quintic not all such choices are independent and one has to divide out by the automorphism group of the ambient space. The resulting space is in general a quasi-projective variety. Furthermore, in order to obtain a smooth Calabi-Yau manifold the $a_{i_1 i_2 \cdots i_5}$ have to be chosen such that P and $\frac{\partial P}{\partial x_j}$ have no common zero (for all j), as this would correspond to a singular Calabi-Yau. The set of all choices of the coefficients $a_{i_1 i_2 \cdots i_5}$ which correspond to such singular manifolds is denoted as the *discriminant locus* of the family of Calabi-Yau spaces associated with P. The discriminant locus forms a complex codimension one subspace of the complex structure moduli space. We present a schematic picture of this in figure 3.1 where the singular divisors are depicted in red.

Among the singular loci in the Calabi-Yau moduli space there exist so called *conifold singularities* which are universal, i.e. they are present in the moduli space of all Calabi-Yau manifolds. They are characterized by the shrinking of a three-sphere S^3 to zero size. Locally, at the singular locus in the complex structure moduli space, the conifold is a cone with sections having the topology of $S^2 \times S^3$ and shrinking to zero size at the origin. Deforming the complex structure away from the singular locus, the S^3 at the tip of the cone remains of finite size and the space is denoted by the term *deformed conifold*.

Kähler moduli space

In this section we will show how the parameter space for the Kähler class turns out to be a Kähler manifold by itself with a holomorphic prepotential analogous to the one of the complex structure moduli space.

Let us define the metric of Kähler deformations as a pairing between (1,1) cohomology classes

$$G(\rho,\sigma) = \frac{1}{2V} \int_X \varrho_{i\bar{j}} \sigma_{k\bar{l}} g^{i\bar{l}} g^{k\bar{j}} g^{\frac{1}{2}} d^6x = \frac{1}{2V} \int_X \varrho \wedge *\sigma, \quad (3.1.43)$$

where V is the volume and we interpret the (1,1)-forms σ and ϱ as metric variations of mixed type $\delta g_{i\bar{j}}$. The above metric is positive definite and symmetric due to the positivity of $g^{i\bar{j}}$ and the symmetry of the pairing. As was observed by Strominger in [23] $G(\varrho,\sigma)$

may be expressed entirely in terms of triple intersection numbers

$$\kappa(\varrho, \sigma, \tau) := \int \varrho \wedge \sigma \wedge \tau, \qquad (3.1.44)$$

giving

$$G(\varrho, \sigma) = -3 \left(\frac{\kappa(\varrho, \sigma, \omega)}{\kappa(\omega, \omega, \omega)} - \frac{3}{2} \frac{\kappa(\varrho, \omega, \omega)\kappa(\sigma, \omega, \omega)}{\kappa^2(\omega, \omega, \omega)} \right). \qquad (3.1.45)$$

Here ω is the Kähler form of X and we have $\kappa(\omega, \omega, \omega) = 6V$. The main ingredient in the derivation of this formula is the identity

$$*\sigma = -\omega \wedge \sigma + \frac{3}{2} \frac{\kappa(\sigma, \omega, \omega)}{\kappa(\omega, \omega, \omega)} \omega \wedge \omega, \qquad (3.1.46)$$

which one can prove as follows. There exists a direct sum decomposition

$$\sigma = \sigma_P + k\omega, \qquad (3.1.47)$$

where σ_P is the *primitive* part of σ and k is a constant. On Kähler manifolds the primitive part of a $(1,1)$-form is defined by the relation $\int_X \omega \wedge *\sigma_P = 0$, i.e. the adjoint of the *Lefschetz operator* (being defined as multiplication by ω) applied to σ_P is zero. A theorem in complex geometry [22] says that for primitive $(1,1)$-forms $*\sigma_P = -\omega \wedge \sigma_P$. One can use this theorem together with the decomposition (3.1.47) to compute

$$\begin{aligned} *\sigma &= *\sigma_P + k * \omega \\ &= -\omega \wedge \sigma_P + \frac{k}{2} \omega^2 \\ \Rightarrow \int \omega \wedge *\sigma &= -\underbrace{\int \omega \wedge \omega \wedge \sigma_P}_{=0} + \frac{k}{2} \int \omega \wedge \omega \wedge \omega. \end{aligned} \qquad (3.1.48)$$

On the other hand we have $\int \omega \wedge *\sigma = \int *\omega \wedge \sigma = \frac{1}{2} \int \omega \wedge \omega \wedge \sigma$, which gives $k = \frac{\int \omega \wedge \omega \wedge \sigma}{\int \omega \wedge \omega \wedge \omega}$. Thus, we have

$$\begin{aligned} *\sigma &= -\omega \wedge \sigma_P + \frac{1}{2} k\omega \wedge \omega - k\omega \wedge \omega + k\omega \wedge \omega, \\ &= (3.1.46). \end{aligned} \qquad (3.1.49)$$

Next, we show how the metric is derived from a Kähler potential. Fix a basis e_A, $A = 1, \cdots, b_2$ of $H^2(X, \mathbb{Z})$ and write the B-field together with the Kähler form in terms of this basis

$$B + i\omega = w^A e_A, \quad w^A = u^A + iv^A. \qquad (3.1.50)$$

A straightforward calculation then yields

$$G_{A\overline{B}} := \frac{1}{2} G(e_A, e_B) = -\frac{\partial}{\partial w^A} \frac{\partial}{\partial w^B} \log \kappa(\omega, \omega, \omega). \qquad (3.1.51)$$

3.1. THE BACKGROUND GEOMETRY

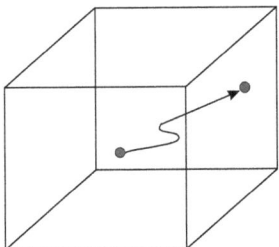

Figure 3.2: The Kähler structure moduli space

Thus, the exponential of the Kähler potential is up to a proportionality factor the volume of the Calabi-Yau manifold. We can even rewrite this Kähler potential further. Define a holomorphic function

$$F(w) = -\frac{1}{3!}\frac{\kappa_{ABC}w^A w^B w^C}{w^0}, \qquad (3.1.52)$$

where $\kappa_{ABC} = \kappa(e_A, e_B, e_C)$ and w^0 has been introduced in order to make F a homogeneous function of degree 2. Introducing affine coordinates $t^A = \frac{w^A}{w^0}$ this can be rewritten as

$$F(t) = -\frac{1}{3!}(w^0)^2 \kappa_{ABC} t^A t^B t^C. \qquad (3.1.53)$$

Taking 3rd derivatives of $F(t)$ we see that κ_{ABC} can be written analogously to C_{ijk} in equation (3.1.39). After some algebra we arrive at the identity

$$\exp(-K) = \frac{4}{3}\int_X \omega^3 = -i\left(w^j \frac{\partial \overline{F}}{\partial \overline{w}^j} - \overline{w}^j \frac{\partial F}{\partial w^j}\right), \qquad (3.1.54)$$

which shows that the moduli space of Kähler deformations is again a Kähler manifold with holomorphic prepotential.

Having described the local situation we next turn to the global description in a short exposition. From the Kähler metric $g_{i\bar{j}}$ one constructs the Kähler form $\omega = ig_{i\bar{j}}dx^i \wedge dx^{\bar{j}}$. It can be shown using the positivity of the volume form

$$\int_X \omega^r > 0, \quad r = \dim_{\mathbb{C}} X, \qquad (3.1.55)$$

that the set of allowed ω's forms a cone known as the *Kähler cone* of X. In string theory one has to add the antisymmetric tensor field $B = B_{i\bar{j}}$ to ω yielding the *complexified* Kähler form $B + i\omega$. Choices of this form which correspond to points on the walls of the Kähler cone correspond to singular manifolds where the condition (3.1.55) is no longer satisfied. Geometrically, one can think of such singularities as a shrinking of a complex curve to zero size within the Calabi-Yau. We sketch this picture of the Kähler moduli space as a bounded domain with its boundary corresponding to the singular geometries (red colour) in figure 3.2.

From the point of view of the Kähler moduli space the conifold singularities of the previous section are also present and lie on the walls of the Kähler cone. Moving the Kähler form away from the wall means introducing a finite S^2 at the tip of the cone. The smooth space obtained this way is denoted by *resolved conifold*.

3.2 Supersymmetric nonlinear Sigma Models

In this section we will review $N = (2, 2)$ nonlinear sigma models and their superconformal algebra. This will be important for the later introduction of twisting to form the so called *topological field theories*. Then we change the view point to the one of the gauged linear sigma model. We will see that this represents a far more general setup which contains the nonlinear sigma model as a specific *phase* in its low energy effective theory.

3.2.1 $N = (2, 2)$ nonlinear Sigma Models

The $N = (2, 2)$ superconformal algebra

The perturbative description of superstring theory involves a two-dimensional superconformal field theory with a central charge of $c = 15$. The most interesting setting for us is the realization of the central charge via $M_4 \times \{\text{an } N = 2, c = 9 \text{ superconformal field theory}\}$, where M_4 denotes Minkowski space, i.e. the $c = 6$ superconformal field theory of four free bosons and their super partners. In our presentation we shall focus on $N = 2, c = 9$ theories as these are the theories describing the dynamics of the superstring on the internal 6-dimensional space. References [26] [27] contain a more detailed version of the material presented here, for a general introduction to conformal field theory see [28].

As a first step we write down the superconformal algebra which forms the backbone of type II perturbative string theory and its twisted topological versions.

$$\begin{aligned}
[L_m, L_n] &= (m-n)L_{m+n} + \frac{c}{12}m(m^2 - 1)\delta_{m+n,0}, \\
[J_m, J_n] &= \frac{c}{3}m\delta_{m+n,0}, \\
[L_n, J_m] &= -mJ_{m+n}, \\
[L_n, G_r^\pm] &= (\frac{n}{2} - r)G_{r+n}^\pm, \\
[J_n, G_r^\pm] &= \pm G_{r+n}^\pm, \\
\{G_r^+, G_s^-\} &= 2L_{r+s} + (r-s)J_{r+s} + \frac{c}{3}(r^2 - \frac{1}{4})\delta_{r+s,0}.
\end{aligned} \qquad (3.2.1)$$

Here, L_n are the familiar generators of the $N = 0$ conformal algebra and $G_n^\pm$ are the modes of the worldsheet super partners $G^\pm(z)$ of the energy-momentum tensor $T(z)$. $G^\pm(z)$ are primary fields of weight $\frac{3}{2}$ and we have two of them as the theory exhibits $N = 2$ supersymmetry. The J_n are the modes of the $U(1)$-current generator $J(z)$ which

3.2. SUPERSYMMETRIC NONLINEAR SIGMA MODELS

is a primary field of weight one. Furthermore, c is the central charge of the algebra and must be equal to 9 if the CFT represents the internal manifold M. In fact in this case c is divisible by 3 and we have $d := \dim_{\mathbb{C}} M = \frac{c}{3}$. The above formulas are valid for sectors where r, s run over half-integral values, denoted by NS sector, and for the case when they take only integral values, denoted by R sector. The algebra (3.2.1) is doubled in type II string theory with one set describing the left moving sector and the other describing the right moving sector. We shall write generators of the right moving sector with barred letters. Let us define fields $T(z) = \sum_{n \in \mathbb{Z}} \frac{L_n}{z^{n+2}}$ with conformal dimension and $U(1)$ charge $(h, Q) = (2, 0)$, a $U(1)$ current $J(z) = \sum_{n \in \mathbb{Z}} \frac{J_n}{z^{n+1}}$ with $(h, Q) = (1, 0)$ and $G^{\pm} = \sum_{r \in \mathbb{Z} \pm \nu} \frac{G_r^{\pm}}{z^{r+\frac{3}{2}}}$ with $(h, Q) = (\frac{3}{2}, \pm 1)$ and ν being 0 or $\frac{1}{2}$. With these definitions equations (3.2.1) can be deduced from the following short distance operator expansions

$$T(z)T(0) = \frac{c}{2z^4} + \frac{2}{z^2}T(0) + \frac{1}{z}\partial T(0),$$

$$T(z)G^{\pm}(0) \sim \frac{3}{2z^2}G^{\pm}(0) + \frac{1}{z}\partial G^{\pm}(0),$$

$$T(z)J(0) \sim \frac{1}{z^2}J(0) + \frac{1}{z}\partial J(0),$$

$$G^{+}(z)G^{-}(0) \sim \frac{2c}{3z^3} + \frac{2}{z^2}J(0) + \frac{2}{z}T(0) + \frac{1}{z}\partial J(0),$$

$$G^{+}(z)G^{+}(z) \sim G^{-}(z)G^{-}(0) \sim 0,$$

$$J(z)G^{\pm}(0) \sim \pm \frac{1}{z}G^{\pm}(0),$$

$$J(z)J(0) \sim \frac{c}{3z^2}. \tag{3.2.2}$$

Let us now pass over to some definitions and the structure of the ground states of the algebra. We denote a *left-chiral* state as a state in the NS Hilbert space satisfying

$$G^{+}_{-\frac{1}{2}}|\phi\rangle = 0, \tag{3.2.3}$$

and an *anti-chiral* state as a state satisfying equation (3.2.3) with G^{+} replaced with G^{-}. Right-chiral states are defined similarly, by replacing G with $\overline{G}$. From now on we shall concentrate on the left-moving sector. States which satisfy in addition to (3.2.3) the condition

$$G^{-}_{n+\frac{1}{2}}|\phi\rangle = G^{+}_{n+\frac{1}{2}}|\phi\rangle = 0 \quad \text{for } n \geq 0 \tag{3.2.4}$$

are called *primary* chiral states. One can write an immediate property of such states by using the commutation relation of the G's

$$\{G^{-}_{\frac{1}{2}}, G^{+}_{\frac{1}{2}}\}|\phi\rangle = (2L_0 - J_0)|\phi\rangle = 0, \tag{3.2.5}$$

which implies the relation $h = \frac{Q}{2}$ for the charge Q and weight h of such a state. For anti-chiral states the analogous equation is $h = -\frac{Q}{2}$. Note that $\{G^{-}_{\frac{1}{2}}, G^{+}_{\frac{1}{2}}\}$ is a positive operator (as $G^+_{\frac{1}{2}} = G^{-\dagger}_{\frac{1}{2}}$), so we have

$$\langle\psi|\{G^-_{\frac{1}{2}}, G^+_{\frac{1}{2}}\}|\psi\rangle \geq 0 \qquad (3.2.6)$$

for any state ψ in the Hilbert space, i.e. $h \geq |Q|/2$. One can show that this inequality is saturated if and only if one is dealing with a primary chiral or anti-chiral state. Furthermore, one can show by using the positivity of the operator

$$\{G^-_{\frac{3}{2}}, G^+_{-\frac{3}{2}}\} = 2L_0 - 3J_0 + 2\frac{c}{3} \qquad (3.2.7)$$

and the property $h = Q/2$ that primary chiral states always satisfy $h \leq c/6$. Up to now, we were solely dealing with states in the NS sector. However, these can be related to states in the R sector by the so called *spectral flow*

$$L_n \mapsto L'_n = L_n + \theta J_n + \frac{1}{6}\theta^2 c\delta_{n,0} \qquad (3.2.8)$$

$$J_n \mapsto J'_n = J_n + \frac{1}{3}\theta c\delta_{n,0} \qquad (3.2.9)$$

$$G^\pm_r \mapsto = (G^\pm_r)' = G^\pm_{r\pm\theta}. \qquad (3.2.10)$$

The above is an algebra automorphism interpolating for $\theta \in \mathbb{Z} + \frac{1}{2}$ between the NS and R sectors, and for $\theta \in \mathbb{Z}$ it maps the NS to NS and R to R. For $\theta = \frac{1}{2}$ one can check that chiral primary states are mapped to the ground states of the Ramond sector, defined by $G^\pm_0|\tilde\phi\rangle = 0$. The anticommutator $\{G^-_0, G^+_0\}$ constrains the conformal weight of these states to satisfy $h = c/24$ and once again one deduces from its positivity that all states in the R sector have $h \geq c/24$. Consider now the flow from the NS sector to the NS sector with flow parameter $\theta = 1$. As the flow is possible for the left and right sectors independently, we choose parameters $(\theta_L, \theta_R) = (1, 0)$. Using this map and starting with (c, c) (here c stands for chiral) primary states in the (NS,NS) sector, we end up with (a, c) (where a stands for anti-chiral) elements again in the (NS, NS) sector. This can be seen immediately form (3.2.8) as $G^+_{-\frac{1}{2}}$ maps to $G^+_{\frac{1}{2}}$ and $G^-_{\frac{1}{2}}$ maps to $G^-_{-\frac{1}{2}}$. We can be even more precise. The vacuum of the left moving chiral sector is a state $|\rho\rangle$ annihilated by $G^\pm_{-\frac{1}{2}}$. Under spectral flow it is mapped to a state $|\tilde\rho\rangle$ given by

$$G^+_{n+\frac{1}{2}}|\tilde\rho\rangle = G^-_{n-\frac{3}{2}}|\tilde\rho\rangle = 0 \text{ for all } n \geq 0. \qquad (3.2.11)$$

Thus, in particular, it is annihilated by the operator

$$\{G^-_{-\frac{3}{2}}, G^+_{\frac{3}{2}}\} = 2L_0 + 3J_0 + 2c/3. \qquad (3.2.12)$$

Using the property $h = -Q/2$ for anti-chiral primary states we deduce that the vacuum of the left moving chiral sector is mapped to the highest weight state of the left moving anti-chiral sector. These properties will become important later when we come to mirror symmetry and analyze the relation between chiral primary fields and cohomology elements of the internal manifold.

3.2. SUPERSYMMETRIC NONLINEAR SIGMA MODELS

The Sigma Model

$N = 2$ superconformal non-linear sigma models are given by a map

$$X : \Sigma \to M, \tag{3.2.13}$$

from the worldsheet Σ being a Riemann surface to the target space M which is a curved Riemannian manifold with non-trivial metric. The bosons X can be thought of as coordinates on the target space and their fermionic superpartners will be sections of the pullback of the tangent bundle of the target space, i.e. $\Psi_+^\mu \in \Gamma(\overline{K}^{\frac{1}{2}} \otimes X^*(\mathcal{T}_M))$ and $\Psi_-^\mu \in \Gamma(K^{\frac{1}{2}} \otimes X^*(\mathcal{T}_M))$ where Γ denotes sections of the indicated bundles. Here, K and $\overline{K}$ denote the canonical and anti-canonical line bundles of Σ (the bundles of one forms of types $(1,0)$ and $(0,1)$, respectively), and $K^{\frac{1}{2}}$ and $\overline{K}^{\frac{1}{2}}$ are square roots of these. The action for such a theory is given by

$$S = 2t \int_\Sigma d^2z \left(\frac{1}{2} g_{\mu\nu}(X) \partial_z X^\mu \partial_{\bar{z}} X^\nu + g_{\mu\nu}(\Psi_+^\mu D_{\bar{z}} \Psi_+^\nu + \Psi_-^\mu D_z \Psi_-^\nu) + \frac{1}{4} R_{\mu\nu\rho\sigma} \Psi_+^\mu \Psi_+^\nu \Psi_-^\rho \Psi_-^\sigma \right),$$
$$(3.2.14)$$

where t is the coupling constant of the theory and $g_{\mu\nu}$ is the metric on the target manifold with $R_{\mu\nu\rho\sigma}$ being its Riemann tensor. Let us now look at the conditions under which this theory has $(2,2)$ superconformal symmetry. As was shown in [29] the condition for $(2,2)$ supersymmetry is that the target manifold be a complex Kähler manifold. To see this, note that the $N=2$ superspace version of (3.2.14) is:

$$S = 2t \int d^2z d^4\theta K(X^i, X^{\bar{j}}), \tag{3.2.15}$$

where the X^i are chiral superfields whose lowest components are the bosonic coordinates above and

$$g_{i\bar{j}} = \frac{2i}{\pi} \frac{\partial^2 K}{\partial X^i \partial X^{\bar{j}}}. \tag{3.2.16}$$

Conformal invariance is an even more stringent condition. Viewing the metric g as a coupling "constant" of the two-dimensional theory one can compute its β function. The result is that the β function, to lowest order, is proportional to the Ricci tensor of the target manifold. Therefore, we see that in order to establish conformal invariance the target manifold has to be chosen to be Ricci flat which by our analysis in section (3.1) is equivalent to a Calabi-Yau manifold.

From now on we shall only deal with sigma models on Kähler manifolds and will therefore rewrite the action (3.2.14) in terms of complexified coordinates. Local complex coordinates on M will be denoted by X^i and their complex conjugates are $X^{\bar{i}}$. The projections of Ψ_+ in $K^{\frac{1}{2}} \otimes X^*(TM^{(1,0)})$ and $K^{\frac{1}{2}} \otimes X^*(T^{(0,1)})$, respectively, will be written as Ψ_+^i and $\Psi_+^{\bar{i}}$. Analogously, Ψ_-^i is the projection of Ψ_- in $\overline{K}^{\frac{1}{2}} \otimes X^*(TM^{(1,0)})$ and $\Psi_-^{\bar{i}}$ is the projection in $\overline{K}^{\frac{1}{2}} \otimes X^*(TM^{(0,1)})$. The action is now given by

$$S = 2t \int_\Sigma d^2z \left(\frac{1}{2} g_{i\bar{j}} \partial_z X^i \partial_{\bar{z}} X^{\bar{j}} + i g_{\bar{i}i} \Psi_-^{\bar{i}} D_z \Psi_-^i + i g_{\bar{i}i} \Psi_+^{\bar{i}} D_{\bar{z}} \Psi_+^i + R_{i\bar{i}j\bar{j}} \Psi_+^i \Psi_+^{\bar{i}} \Psi_-^j \Psi_-^{\bar{j}} \right),$$
$$(3.2.17)$$

where t denotes an arbitrary coupling constant and we have the connection

$$\begin{aligned}D_{\bar{z}}\Psi^i_+ &= \partial_{\bar{z}}\Psi^i_+ + \Gamma^i_{kl}\partial_{\bar{z}}X^k\Psi^l_+, \\ D_z\Psi^i_- &= \partial_z\Psi^i_- + \Gamma^i_{kl}\partial_z X^k\Psi^l_-.\end{aligned} \quad (3.2.18)$$

The supersymmetry variations of the fields in this action are parametrized in terms of infinitesimal fermionic parameters ϵ_-, $\bar{\epsilon}_-$ (being holomorphic sections of $K^{-\frac{1}{2}}$) and ϵ_+, $\bar{\epsilon}_+$ (being antiholomorphic sections of $\overline{K}^{-\frac{1}{2}}$)

$$\begin{aligned}\delta X^i &= -\epsilon_-\Psi^i_+ + \epsilon_+\Psi^i_- \\ \delta X^{\bar{i}} &= \bar{\epsilon}_-\Psi^{\bar{i}}_+ - \bar{\epsilon}_+\Psi^{\bar{j}}_- \\ \delta\Psi^i_+ &= 2i\bar{\epsilon}_-\partial_z X^i + \epsilon_+\Psi^j_+\Gamma^i_{jm}\Psi^m_- \\ \delta\Psi^{\bar{i}}_+ &= -2i\epsilon_-\partial_z X^{\bar{i}} + \bar{\epsilon}_+\Psi^{\bar{j}}_+\Gamma^{\bar{i}}_{\bar{j}\bar{m}}\Psi^{\bar{m}}_- \\ \delta\Psi^i_- &= -2i\bar{\epsilon}_+\partial_{\bar{z}}X^i + \epsilon_-\Psi^j_+\Gamma^i_{jm}\Psi^m_- \\ \delta\Psi^{\bar{i}}_- &= 2i\epsilon_+\partial_{\bar{z}}X^{\bar{i}} + \bar{\epsilon}_-\Psi^{\bar{j}}_-\Gamma^{\bar{i}}_{\bar{j}\bar{m}}\Psi^{\bar{m}}_+.\end{aligned} \quad (3.2.19)$$

Apart from variation under supersymmetry the action is also invariant under the so called R-symmetries $U(1)_{R/L}$ which can be identified with the left moving and right moving symmetries under J and $\bar{J}$ of the $N = (2,2)$ superconformal algebra (3.2.1). For our purposes it is convenient to define $U(1)$ generators which are linear combinations of these, namely F_V corresponding to the vector R-symmetry $U(1)_V = U(1)_L + U(1)_R$ and F_A corresponding to the axial R-symmetry $U(1)_A = U(1)_L - U(1)_R$. On the quantum level $U(1)_V$ remains a symmetry of the theory while the $U(1)_A$ R-symmetry is broken to $\mathbb{Z}_{2k}$ where k is

$$k = \int_\Sigma c_1(X^*(TM^{(1,0)})) = \int_\Sigma X^* c_1(TM^{(1,0)}) = \langle c_1(M), X_*[\Sigma]\rangle. \quad (3.2.20)$$

Thus, we see that $U(1)_A$ is not anomalous if and only if $c_1(M) = 0$, namely when M is a Calabi-Yau manifold.

Let us also introduce generators $Q_\mp$ and $\overline{Q}_\mp$ corresponding to the supersymmetry transformations generated by $\epsilon^\pm$ and $\bar{\epsilon}^\pm$. From the conformal field theory point of view these are obtained by the contour integrations

$$Q_+ = \oint G^-, \quad Q_- = \oint \overline{G}^-, \quad \overline{Q}_+ = \oint G^+, \quad \overline{Q}_- = \oint \overline{G}^+. \quad (3.2.21)$$

Now we are ready to write down the algebra which the conserved charges of the field theory defined by the action (3.2.17) fulfill

$$\begin{aligned}&Q_+^2 = Q_-^2 = \overline{Q}_+^2 = \overline{Q}_-^2 = 0 \\ &\{Q_\pm, \overline{Q}_\pm\} = H \pm P, \quad \{\overline{Q}_+, \overline{Q}_-\} = \{Q_+, Q_-\} = \{Q_-, \overline{Q}_+\} = \{Q_+, \overline{Q}_-\} = 0, \\ &[M_E, Q_\pm] = \mp Q_\pm, \quad [M_E, \overline{Q}_\pm] = \mp \overline{Q}_\pm, \\ &[F_V, Q_\pm] = -Q_\pm, \quad [F_V, \overline{Q}_\pm] = \overline{Q}_\pm, \\ &[F_A, Q_\pm] = \mp Q_\pm, \quad [F_A, \overline{Q}_\pm] = \pm \overline{Q}_\pm.\end{aligned} \quad (3.2.22)$$

3.2. SUPERSYMMETRIC NONLINEAR SIGMA MODELS

Here M_E is the generator of the compact Euclidean rotation group $U(1)_E$ obtained after Wick rotation from the two dimensional Lorentz group $SO(1,1)$. Furthermore, beside the supersymmetry generators, one has the generator of (euclidian) time translations H, and the generator of translations P. The above operators act on operators $\mathcal{O}_\phi$ corresponding to a field ϕ by $[Q, \mathcal{O}_\phi] = \delta_Q \mathcal{O}_\phi$, where Q is a general operator and $\delta_Q \mathcal{O}_\phi$ describes the infinitesimal field transformation.

Deformations

Perturbations of two-dimensional conformal field theory are parametrized by marginal operators of weight $h+\bar{h} = 2$. Of most interest to us are operators which do not change the central charge of a given conformal field theory and so can be used to deform the original theory to as "nearby conformal field theory". Such operators are spinless operators with $h = \bar{h} = 1$. In order for these operators to remain of type $(1,1)$ even after deformation of the theory, they have to be "truly marginal". It can be shown [30] [31] that such operators are given by the following constructions:

- Take a field ϕ in the ring (c, c) with $h = \bar{h} = \frac{1}{2}, Q = \bar{Q} = 1$. Then define $\hat{\phi}$ by

$$\hat{\phi}(w, \bar{w}) \equiv \oint dz G^-(z) \phi(w, \bar{w})). \quad (3.2.23)$$

$\hat{\phi}$ has $h = \frac{1}{2} + \frac{1}{2} = 1$ and $\bar{h} = \frac{1}{2}$, furthermore $Q = 0$ and $\bar{Q} = 1$. Now define

$$\Phi_{(1,1)}(w, \bar{w}) \equiv \oint d\bar{z}\overline{G}^-(\bar{z}) \hat{\phi}(w, \bar{w}). \quad (3.2.24)$$

It follows immediately that $\Phi_{(1,1)}$ has $h = \bar{h} = 1, Q = \bar{Q} = 0$. It is a truly marginal operator.

- Start with $\phi \in (a, c)$ with $h = \bar{h} = \frac{1}{2}$ and $Q = -\bar{Q} = 1$. Next, define

$$\hat{\phi}(w, \bar{w}) \equiv \oint dz \overline{G}^-(\bar{z}) \phi(w, \bar{w}), \quad (3.2.25)$$

and

$$\Phi_{(-1,1)} \equiv \oint dz G^+(z) \hat{\phi}(w, \bar{w}) = (G^+_{-\frac{1}{2}} \overline{G}^-_{-\frac{1}{2}} \phi)(w, \bar{w}). \quad (3.2.26)$$

Its definition shows that $\Phi_{(-1,1)}$ has $h = \bar{h} = 1$, $Q = \bar{Q} = 0$. This operator is truly marginal.

These operators have the following interpretation from the point of view of the nonlinear sigma model. As one can show (a, c) fields can be written as $b_{i\bar{j}} \Psi^i_+ \Psi^{\bar{j}}_-$ with $b_{i\bar{j}}$ a harmonic $(1, 1)$-form on M. The map between (a, c) fields and marginal operators applied to this field is to lowest order $b_{i\bar{j}} \partial_z X^i \partial_{\bar{z}} X^{\bar{j}}$. This shows, that in this case the marginal operator corresponds to deformations of the Kähler class of M. On the other hand deformations arising from (c, c) fields lead to pure-index type metric perturbations and thus to complex structure deformations of M.

3.2.2 Linear Sigma Model view point

In [32] Witten shows how the nonlinear sigma model described in the previous section can be understood as a specific phase in a more general theory, namely the so called gauged linear sigma model. This view point turns out to be very powerful as the linear sigma model incorporates a rich phase structure including points where a Landau-Ginzburg description arises. This way the important CY/LG correspondence can be deduced from an underlying theory. Way shall describe this construction briefly here.

Let us consider the simplest setup, where we have a two-dimensional $N = 2$ linear sigma model which is gauged under the group $U(1)$, and has thus a single vector superfield V. Solving for the auxiliary fields by their equations of motion, one gets

$$D = -e^2 \left(\sum_i Q_i |\phi_i|^2 - r \right)$$
$$F_i = \frac{\partial W}{\partial \phi_i}, \qquad (3.2.27)$$

where D is the D-term, W is the superpotential, ϕ_i are scalar components of chiral superfields with charge Q_i and r is the Fayet-Iliopolous parameter. These fields sit together with the scalar fields ϕ_i, σ in the scalar potential

$$U(\phi_i, \sigma) = \frac{1}{2e^2} D^2 + \sum_i |F_i|^2 + 2\bar{\sigma}\sigma \sum_i Q_i^2 |\phi_i|^2. \qquad (3.2.28)$$

Here, Q_i is the charge under $U(1)$ corresponding to the field ϕ_i. Next, we choose n of the chiral superfields Φ_i to have charge 1 and denote them by S_i, and one field P of charge $-n$. Then, one can write down the following gauge invariant superpotential

$$W = P \cdot G(S_1, \cdots, S_n), \qquad (3.2.29)$$

where G is a homogeneous polynomial of degree n and we will think of the S_i as homogeneous variables corresponding to $\mathbb{P}^{n-1}$. G has furthermore to be chosen such that the equations

$$0 = \frac{\partial G}{\partial S_1} = \cdots = \frac{\partial G}{\partial S_n} \qquad (3.2.30)$$

have no solutions except at $S_i = 0$ (this fact is called transversality of G). Geometrically this means that $G = 0$ cuts out a smooth hypersurface X in $\mathbb{P}^{n-1}$. Having specified the setup we conclude that the scalar potential (3.2.28) can be expressed in terms of s_i, p, σ and the homogeneous polynomial G as

$$U = |G(s_i)|^2 + |p|^2 \sum_i |\frac{\partial G}{\partial s_i}|^2 + \frac{1}{2e^2} D^2 + 2|\sigma|^2 \left(\sum_i |s_i|^2 + n^2 |p|^2 \right), \qquad (3.2.31)$$

with

$$D = -e^2 \left(\sum_i \bar{s}_i s_i - n\bar{p}p - r \right). \qquad (3.2.32)$$

3.2. SUPERSYMMETRIC NONLINEAR SIGMA MODELS

Now we are ready to present the two most important phases for two extremal values of r.

The Calabi-Yau phase

This picture arises as the low energy physics of the phase $r \gg 0$. Minimizing the scalar potential requires $D = 0$ which in turn implies that not all s_i can vanish. As a consequence we see that the vanishing of $|p|^2 \sum_i |\partial_i G|^2$ in the potential requires that $p = 0$. Therefore, vanishing of D leads us to the equation

$$\sum_i \bar{s}_i s_i = r. \tag{3.2.33}$$

As we are dealing with a gauge theory the space of solutions of (3.2.33) must be divided by the gauge group $U(1)$ and this way we arrive at the complex projective space $\mathbb{P}^{n-1}$, with Kähler class proportional to r. Finally, the condition $G = 0$ has to be imposed on the vacuum to ensure the vanishing of the first term in the expression for the scalar potential, and for the vanishing of the last term σ has also to be set to zero. We see that the space of classical vacua is isomorphic to the hypersurface $X \subset \mathbb{P}^{n-1}$ defined by $G = 0$. In order for the hypersurface to be Calabi-Yau in $\mathbb{P}^{n-1}$ the degree of the defining equation must be n. This restriction is from the point of view of the linear sigma model not accidental as it is automatically chosen in order to ensure anomaly-free R-invariance. All modes which are not parallel to oscillations tangent to X have masses at tree level. Therefore, the low energy theory is precisely a sigma model with target space X and Kähler class proportional to r.

The Landau-Ginzburg phase

Next we turn to the phase $r \ll 0$. First of all we see that the vanishing of D requires $p \neq 0$. Then the vanishing of $|p|^2 \sum_i |\partial_i G|^2$ leads to the condition $s_i = 0$ for all i, where we have used the transversality of G. Having established this, the modulus of p must be $|p| = \sqrt{-r/n}$ which shows that the theory has a unique classical vacuum up to gauge transformation. A property of this vacuum is that the s_i remain massless in expansions around it (this holds for $n \geq 3$). Integrating out the massive field p is equivalent to setting p to its expectation value. This way one arrives at the effective superpotential $\tilde{W} = \sqrt{-r} \cdot W(s_i)$. This effective superpotential describes a Landau-Ginzburg theory as it has a unique classical vacuum with a degenerate critical point at the origin. Indeed it is even a Landau-Ginzburg orbifold, the reason being that the vacuum expectation value of p does not completely destroy the gauge invariance but rather breaks it to the discrete group with the action

$$s_i \to \xi s_i, \tag{3.2.34}$$

where ξ is an nth root of unity. This residual gauge invariance means that we are dealing with a $\mathbb{Z}_n$ orbifold of a Landau-Ginzburg theory.

Let us look at the consequences of this observation. Allowing renormalization group flow the Landau-Ginzburg theory will reach a conformally invariant infrared fixed point.

Indeed, one can show that there is an isomorphism between a conformal field theory minimal model at level P and the Landau-Ginzburg theory of a single chiral superfield X with superpotential $W = X^{P+2}$. Therefore, one immediately learns that a Landau-Ginzburg theory with scalar action

$$S = \int d^2z d^4\theta \sum_{j=1}^n K_j(S_j, \bar{S}_j) + \left(d^2z d^2\theta \sum_{j=1}^n S_j^{P_j+2} + h.c. \right) \tag{3.2.35}$$

is isomorphic to the tensor product $\otimes_{j=1}^n MM_{P_j}$ of minimal model at level P_j at its conformally invariant fixed point.

Consequences

Here we want to outline the overall picture which arises from the phase structure of the linear sigma model. First of all note that one should take $|r|$ to be large in each regime in order to suppress the massive excitations which would cause significant changes to a non-linear sigma model or a Landau-Ginzburg model. r determines the Kähler form in the Calabi-Yau case and the expectation values of twist fields in the Landau-Ginzburg theory. Hence, we can think of r as a Kähler moduli space parameter and the moduli space will consist of $\mathbb{R}$ divided into two regions $r \geq 0$ and $r \leq 0$. Physically, we have the interpretation of an infinite volume Calabi-Yau space in the one region and a Landau-Ginzburg orbifold point with an enhanced quantum symmetry at the value $r \to \infty$. In fact, we have to complexify the variable r by including the B-field in the way $r \to t = b + ir$. As shifts of b do not affect the theory, the natural complex variable to use is $w = e^{2\pi i (b+ir)}$. This way one can map the sigma model region of the moduli space to the upper hemisphere of a sphere. Similar arguments for the Landau-Ginzburg region [33], [32] lead to a map from the complexified region $r < 0$ to the lower hemisphere of a sphere. The lower hemisphere which gives rise to a Landau-Ginzburg theory can be thought of as the analytic continuation of a Calabi-Yau sigma model with a particular Kähler class. The picture one should have in mind here is depicted in figure (3.3).

The setup described in this section can be generalized to more involved linear sigma models as follows:

- <u>Hypersurfaces in weighted projective space and generalizations:</u> Consider a $U(1)$ gauge theory with n chiral superfields S_i, $i = 1, \cdots, n$, of charge q_i, together with one more chiral superfield P of charge $-\sum_i q_i$. This choice ensures $\sum_i Q_i = 0$, which is equivalent to anomaly-free R symmetry and leads to Calabi-Yau manifolds at low energies. However, as different fields now may have different $U(1)$ charge the D-term vanishing condition tells us that we are dealing with the weighted projective space $\mathbb{P}_{n-1}^{q_1,\cdots,q_n}$ with Kähler class proportional to r. The hypersurface $G = 0$ cuts out a Calabi-Yau in this space.

    Another way to generalize the previous construction is by considering degree d_r polynomials G_r of $S_1, \cdots, S_n (r = 1, \cdots, l)$. Putting appropriate transversality con-

3.2. SUPERSYMMETRIC NONLINEAR SIGMA MODELS

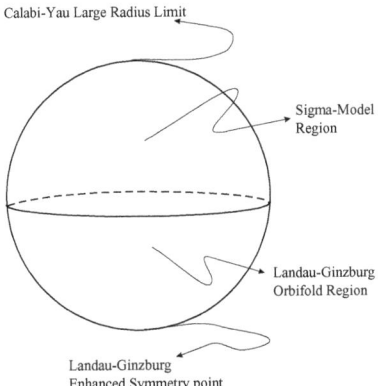

Figure 3.3: The Kähler moduli space associated to the linear sigma model.

ditions on the G_r one sees that the submanifold

$$X = \{G_r(s_1, \cdots, s_n) = 0 \forall r\} \subset \mathbb{P}^{n-1} \qquad (3.2.36)$$

is a smooth complex manifold of dimension $n - l - 1$ which is denoted by the term *complete intersection*. Requiring $n - d_1 - \cdots - d_l = 0$ furthermore ensures that the first chern class of this manifold is 0.

Combining these two generalizations and introducing several $U(1)$ gauge fields one can even describe complete intersections in toric ambient spaces by this construction.

- **Non-compact Calabi-Yau manifolds:** Consider a theory with chiral matter fields $\overline{S_1, \cdots, S_n}$ with charges $\overline{Q_1, \cdots, Q_n}$. Note that this time we do not add a superpotential to the theory. Assuming that there are both positive and negative Q_i's the vacuum manifold is non-compact. Let now $Q_1, \cdots, Q_l$ be positive and $Q_{l+1}, \cdots, Q_n$ be negative. In this case the vacuum manifold is the $U(1)$ quotient of $\sum_{i=1}^{l} Q_i |s_i|^2 = r + \sum_{j=l+1}^{n} |Q_j||s_j|^2$, which can be described as a vector bundle over weighted projective space

$$X = \left[\bigoplus_{j=l+1}^{n} \mathcal{L}^{Q_j} \to \mathbb{P}_{l-1}^{(Q_1, \cdots, Q_l)} \right]. \qquad (3.2.37)$$

- **Hypersurfaces in Grassmannians:** To achieve a realization of a sigma model with target space a Calabi-Yau in a Grassmannian of k planes in complex n space (denoted by $G(k,n)$), one includes kn chiral superfields $S_\lambda^i, i = 1, \cdots, k, \lambda = 1, \cdots, n$.

One can think of the S_λ^i as matrix elements of a $k \times n$ matrix S with an adjoint $\overline{S}$. The group $G = U(k)$ acts on the S's by

$$S_\lambda^i \to M_{i'}^i S_\lambda^{i'}, \qquad (3.2.38)$$

for $M_{i'}^i \in U(k)$. Furthermore, one has to include a $k \times k$ hermitian matrix V of vector superfields and take the Lagrangian to be the gauge and matter kinetic energy plus a Fayet-Iliopoulos term for the central factor $U(1) \subset U(k)$. Next, introduce a complex superfield P which transforms as $P \to (\det M)^{-n} P$ and a superpotential $W = PG(S_\lambda^i)$, where G is a polynomial that transforms as $G \to (\det M)^n G$. This way we obtain a theory which in the regime $r \gg 0$ is described as a sigma model with target space $X \subset G(k, n)$ given by $G = 0$.

We will analyze the topological string on all these three types of spaces in this thesis.

3.3 Twisting the $N = (2,2)$ theories

Twisting nonlinear sigma models turns them into topological field theories which do not depend on the worldsheet metric any more. Also the dependence on target space parameters is reduced to either Kähler or complex structure deformations. States of the topological field theory are isomorphic to ground states of the $N = (2,2)$ CFT forming a vector bundle over the moduli space of deformation parameters.

3.3.1 Generalities about topological field theories

Topological field theories arise from an underlying theory with a nilpotent BRST operator Q. The symmetry generated by Q is used to define physical states which lie in Q-cohomology classes with trivial states being Q-exact. The analogy with De Rham cohomology is not only accidental at this point as in the case of supersymmetric sigma models there is a map between the cohomology classes of the physical states and those of the target space manifold. The identification works as follows. Consider the supersymmetric version of quantum mechanics of a particle moving in a Riemannian manifold M of dimension n and metric g, and supersymmetry generator Q. Due to the relation

$$H = \frac{1}{2}\{Q, Q^\dagger\} \geq 0 \qquad (3.3.1)$$

a state has zero energy if and only if it is annihilated by Q and $Q^\dagger$:

$$H|\alpha\rangle = 0 \Leftrightarrow Q|\alpha\rangle = \overline{Q}|\alpha\rangle = 0. \qquad (3.3.2)$$

Witten showed [34] that the Q-cohomology (i.e. if one takes the BRST operator to be Q) of the theory corresponds to nothing else but the ground states of the theory. His argument goes roughly as follows. If a state $|\alpha\rangle$ has energy E_n and is Q-closed, $Q|\alpha\rangle = 0$, then by the relation $1 = (QQ^\dagger + Q^\dagger Q)/(2E_n)$ we have $|\alpha\rangle = QQ^\dagger|\alpha\rangle/(2E_n)$, and thus $|\alpha\rangle$

3.3. TWISTING THE $N = (2,2)$ THEORIES

is Q-exact. At zero energy this is no longer possible and the cohomology is nothing but the space of the ground states. The space of the ground states is special due to another property as well. Consider the operator $Q_1 := Q + Q^\dagger$ obeying

$$Q_1^2 = 2H. \tag{3.3.3}$$

As the Hamiltonian commutes with the supercharges this operator preserves each energy level and maps the space of even fermionic number $\mathcal{H}_{(n)}^B$ at energy level E_n to the space of odd fermionic number $\mathcal{H}_{(n)}^F$ and vice versa. For $E_n > 0$ we have $Q_1^2 = 2E_n$ at the nth level, and thus Q_1 is invertible and defines an isomorphism

$$\mathcal{H}_{(n)}^B \cong \mathcal{H}_{(n)}^F. \tag{3.3.4}$$

From this we see that the bosonic and fermionic states are paired at each excited level. However, this argument does not go through at the zero energy level $\mathcal{H}_{(0)}$ as the operator Q_1 squares to zero and there is no isomorphism. If we now consider continuous deformations of the theory which preserve supersymmetry the number of states at each energy level will change. But due to the isomorphism (3.3.4) the zero energy level can receive and loose states only in pairs of a bosonic together with a fermionic state. Thus the net number of bosonic minus the number of fermionic ground states is an invariant

$$\dim\mathcal{H}_{(0)}^B - \dim\mathcal{H}_{(0)}^F = \text{Tr}(-1)^F e^{-\beta H}. \tag{3.3.5}$$

Furthermore Witten observed that the Q operator can be identified with the de Rham operator d and one has

$$H = \frac{1}{2}\{Q,\overline{Q}\} = \frac{1}{2}(dd^\dagger + d^\dagger d) = \frac{1}{2}\Delta, \tag{3.3.6}$$

where Δ is the Laplace-Beltrami operator. Therefore, we see that the supersymmetric ground states (i.e. the zero energy states) are simply the harmonic forms

$$\mathcal{H}_{(0)} = \mathcal{H}(M,g) = \bigoplus_{p=0}^{n} \mathcal{H}^p(M,g). \tag{3.3.7}$$

From this it follows that

$$H^p(Q) = H^p_{DR}(M), \tag{3.3.8}$$

where p is the fermion number on the left hand side and the form-degree on the right hand side. The supersymmetric index is the Euler characteristic of the Q-complex, namely

$$\text{Tr}(-1)^F = \sum_{p=0}^{n}(-1)^p \dim H^p(Q) = \sum_{p=0}^{n}(-1)^p \dim H^p_{DR}(M) = \chi(M). \tag{3.3.9}$$

Supersymmetric Localization

In supersymmetric field theories there exists a mechanism known as supersymmetric localization which reduces the path integral to fixed points of the supersymmetric action. In our presentation we follow a path of argumentation first outlined by Witten [35]. Consider an arbitrary quantum field theory with the function space over which one has to integrate denoted by $\mathcal{E}$. Suppose that the theory admits a group of symmetries F which acts free on $\mathcal{E}$. This allows us to write $\mathcal{E}$ as a fibration $\mathcal{E} \to \mathcal{E}/F$, and the path integral can be evaluated by first integrating over the fibers F of this fibration. This reduces the integration space from $\mathcal{E}$ to $\mathcal{E}/F$. For F invariant observables $\mathcal{O}$ the integration over the fibers is particularly simple and just gives a factor of $\mathrm{vol}(F)$ (which is the volume of the group F):

$$\int_{\mathcal{E}} e^{-L} \mathcal{O} = \mathrm{vol}(F) \cdot \int_{\mathcal{E}/F} e^{-L} \mathcal{O}. \qquad (3.3.10)$$

Let us apply this prescription to the case where we have a supersymmetric theory with the supercharge Q being the BRST operator and F is the supergroup generated by it. The special feature of this specific case is that the volume of the group F is zero, since for a fermionic variable θ,

$$\int d\theta \cdot 1 = 0. \qquad (3.3.11)$$

From this and equation (3.3.10) it follows that if Q acts freely, the expectation value of any Q invariant operator $\mathcal{O}$ vanishes.

The general case, however, is that F does not act freely. Let us denote its fixed point locus by $\mathcal{E}_0$. Furthermore, let $\mathcal{C}$ be an F-invariant neighborhood of $\mathcal{E}_0$ and $\mathcal{E}'$ its complement. As the path integral restricted to $\mathcal{E}'$ vanishes by the above argument, the entire contribution to the path integral comes from the integral over $\mathcal{C}$. Note that $\mathcal{C}$ can be an arbitrarily small neighborhood which localizes the computation to an integral on $\mathcal{E}_0$.

For pedagogical reasons we will sketch the way this works in the case of a 0 dimensional quantum field theory, but the proof can be extended to an arbitrary dimension. Consider the path integral $Z := \int dX d\psi_1 d\psi_2 e^{-S(X,\psi_1,\psi_2)}$ with the action given by

$$S(X, \psi_1, \psi_2) := \frac{1}{2}(\partial h)^2 - \partial^2 h \psi_1 \psi_2. \qquad (3.3.12)$$

Here ψ_1, ψ_2 are fermionic Grassmann variables and X is a bosonic coordinate. This action is supersymmetric in the sense that it is invariant under the transformations

$$\begin{aligned}
\delta X &= \epsilon^1 \psi_1 + \epsilon^2 \psi_2, \\
\delta \psi_1 &= \epsilon^2 \partial h, \\
\delta \psi_2 &= -\epsilon^1 \partial h.
\end{aligned} \qquad (3.3.13)$$

The fixed locus of the above variable transformations is given by $\partial h = 0$. Thus we see that we get only contributions from critical points $h'(X_c) = 0$. These contributions can be calculated by considering $h(X) = h(X_c) + \frac{\kappa_c}{2}(x-x_c)^2$, with $\kappa_c = h''(X_c)$. Thus the partition function can be calculated through the Gaussian integral

$$Z = \sum_{X_c} \int \frac{dX d\psi_1 d\psi_2}{\sqrt{2\pi}} e^{-\frac{1}{2}\kappa_c^2(X-X_c)^2 + \kappa_c \psi_1 \psi_2} = \sum_{X_c} \frac{h''(X_c)}{|h''(X_c)|}. \qquad (3.3.14)$$

This result implies that Z counts the sum of zeros of $h'(X)$ weighted with $+1(-1)$ for positive (negative) slope at $h'(X_c)$. Note, that this "index" is invariant under deformations of $h(X)$ as a $+1$ zero of $h'(X)$ can only disappear together with a -1 zero under deformations, which leave the behavior of $h'(X)$ for $|X| \to \infty$ invariant.

3.3.2 The A- and B-twists

In this section we leave the path of general considerations and outline how one can turn $N = (2,2)$ theories into cohomological field theories.

By twisting one means a modification of the Euclidean rotation group $U(1)_E$ by a generator of the global $U(1)$ R-symmetry groups. This amounts to defining a $U(1)_{E'}$ with $M_{E'} = M_E + R$ as the new generator of the Euclidean rotation group. The reason why this twist is performed is that some of the $Q_\pm$ and $\overline{Q}_\pm$ can be made to transform as scalars under $U(1)_{E'}$. As these "scalar" operators are then globally defined on worldsheets of arbitrary genus they can be used to define a cohomological theory on an arbitrary Riemann surface. This would fail in the untwisted case as there are no covariantly constant spinors on a Riemann surface of arbitrary genus.

In the $(2,2)$ theory there are two fundamentally different twists possible,

$$\begin{aligned} \text{A-Twist:} &\quad M_{E'} = M_E + F_V \\ \text{B-Twist:} &\quad M_{E'} = M_E + F_A. \end{aligned} \qquad (3.3.15)$$

This leads to a change of the "spin" of the fields as follows. Consider, as an example, a chiral superfield of trivial R-charges $q_V = q_A = 0$

$$\Phi = X + \theta^+ \Psi_+ + \theta^- \Psi_- + \cdots. \qquad (3.3.16)$$

The zero M_E charge, vector R-charge and axial R-charge of the lowest component X lead to a zero $M_{E'}$ charge which shows that X remains a scalar field after twisting. On the other hand, the M_E charge of Ψ_- is 1 which means that Ψ_- is a spinor field, or a section of the spinor bundle $\sqrt{K}$ before twisting. After A-twist, it has $M_{E'}$ charge $1 + q_V = 0$ and it thus becomes a scalar field. The B-twist gives $M_{E'}$ charge $1 + q_A = 2$ and transforms the spinor into a vector or one-form field being a section of K. We summarize this and the results for other components in table (3.3.1).

Twisting affects the spin of the supercharges as well. From the commutation relations of M_E, F_V, F_A (3.2.22) one can deduce the changes in the transformation rules. This is summarized in table 3.3.2. We see that in the A-twisted theory $\overline{Q}_+$ and Q_- have spin

	Before Twisting				A twist		B twist	
	$U(1)_V$	$U(1)_A$	$U(1)_E$	$\mathcal{L}$	$U(1)'_E$ $\mathcal{L}$		$U(1)'_E$ $\mathcal{L}$	
X	0	0	0	$1_{\mathbb{C}}$	X 0 $1_{\mathbb{C}}$		X 0 $1_{\mathbb{C}}$	
Ψ^i_-	-1	1	1	$K^{\frac{1}{2}}$	χ^i 0 $1_{\mathbb{C}}$		ρ^i_z 2 K	
$\bar\Psi^{\bar i}_+$	1	1	-1	$\bar K^{\frac{1}{2}}$	$\chi^{\bar i}$ 0 $1_{\mathbb{C}}$		$-\frac{1}{2}(\theta^{\bar i}+\eta^{\bar i})$ 0 $1_{\mathbb{C}}$	
$\bar\Psi^{\bar i}_-$	1	-1	1	$K^{\frac{1}{2}}$	$\rho^{\bar i}_z$ 2 K		$\frac{1}{2}(\theta^{\bar i}-\eta^{\bar i})$ 0 $1_{\mathbb{C}}$	
Ψ^i_+	-1	-1	-1	$\bar K^{\frac{1}{2}}$	$\rho^i_{\bar z}$ -2 $\bar K$		$\rho^i_{\bar z}$ -2 $\bar K$	

Table 3.3.1: In this table, $\mathcal{L}$ is the complex line bundle on Σ in which the field takes values. We also indicate the names of the fields in the A and B model.

zero, while in the B-twisted theory $\overline{Q}_+$ and $\overline{Q}_-$ are the spin-zero charges. Thus we notice that the combinations

$$Q_A = \overline{Q}_+ + Q_-$$
$$Q_B = \overline{Q}_+ + \overline{Q}_- \qquad (3.3.17)$$

define scalar, nilpotent operators which can be used to define two different cohomological theories, the topological A- and the topological B-model respectively. We can also call the A-twist a $(+,-)$ twist and the B-twist a $(+,+)$ twist according to the relevant $U(1)$ charges. Let us look at the twisting from the CFT point of view. A $+/-$ twist in the left respectively right moving sector gives

$$\hat T(z) = T(z) \pm' \frac{1}{2}\partial J(z) \rightarrow \hat L_0 = L_0 \pm' \frac{1}{2}J_0. \qquad (3.3.18)$$

This immediately leads to the following short distance expansions

$$\hat T(z)\hat T(0) \sim \frac{2}{z^2}\hat T(0) + \frac{1}{z}\partial\hat T(0),$$
$$\hat T(z)G^\pm(0) \sim \frac{3\pm'\mp 1}{2z^2}G^\pm(0) + \frac{1}{z}\partial G^\pm(0),$$
$$\hat T(z)J(0) \sim \frac{1}{z^2}J(0) + \frac{1}{z^2}\partial J(0) \mp' \frac{c}{3z^3},$$
$$G^+(z)G^-(0) \sim \frac{2c}{3z^3} + \frac{2}{z^2}J(0) + \frac{2}{z}\hat T(0) + \frac{1\mp' 1}{z}\partial J(0). \qquad (3.3.19)$$

Here we stress two points which will become important later on.

- No ghost system is required to quantize the world sheet theory as there is no central term in the first OPE.

- Looking at the third OPE we see that $J(z)$ has an anomalous transformation. In particular we have

$$\nabla^\mu J_\mu = -3\chi(\Sigma_g) \qquad (3.3.20)$$

which is completely analogous to the ghost number anomaly of the bosonic string with ghost current $J_{\text{ghost}} = -:bc:$ in the BRST quantization.

3.3. TWISTING THE N = (2,2) THEORIES

	Before Twisting				A twist		B twist	
	$U(1)_V$	$U(1)_A$	$U(1)_E$	$\mathcal{L}$	$U(1)'_E$	$\mathcal{L}$	$U(1)'_E$	$\mathcal{L}$
Q_-	-1	1	1	$K^{\frac{1}{2}}$	0	$1_\mathbb{C}$	2	K
$\overline{Q}_+$	1	1	-1	$\bar{K}^{\frac{1}{2}}$	0	$1_\mathbb{C}$	0	$1_\mathbb{C}$
$\overline{Q}_-$	1	-1	1	$K^{\frac{1}{2}}$	2	K	0	$1_\mathbb{C}$
Q_+	-1	-1	-1	$\bar{K}^{\frac{1}{2}}$	-2	$\bar{K}$	-2	$\bar{K}$

Table 3.3.2: In this table, $\mathcal{L}$ is the complex line bundle on Σ in which the supercharges take values.

3.3.3 Physical Observables of the topological theories

In a topologically twisted theory one defines physical operators to be operators that commute with $Q = Q_A$ or Q_B. Furthermore, physical states are labeled by Q-cohomology classes, being in one-to-one correspondence with the ground states of the supersymmetric theory. Let us analyze the consequences of this approach.

An operator ϕ is called *chiral* if

$$[Q_B, \phi] = 0. \tag{3.3.21}$$

The lowest component of a chiral superfield Φ obeys $[\overline{Q}_+, \phi] = 0$ which identifies it as a chiral operator. The equations (3.2.21) furthermore show that ϕ belongs to the (c,c)-ring of the superconformal theory. On the other hand an operator ϕ is called *twisted chiral* if

$$[Q_A, \phi] = 0. \tag{3.3.22}$$

The lowest component ν of a twisted chiral superfield Σ (i.e. a superfield satisfying $\overline{D}_+ \Sigma = D_- \Sigma = 0$) obeys $[\overline{Q}_+, \nu] = [Q_-, \nu] = 0$ and is thus a twisted chiral operator. Following the relations (3.2.21) we see that such operators belong to the (a,c) ring.

In the cohomological theory the operators have to fulfill either of the closeness conditions (3.3.21), (3.3.22). Furthermore, operators become equivalent if they are equal up an *exact* operator $d\Lambda = [Q, \Lambda]_\pm$, i.e.

$$\phi \sim \phi + [Q, \Lambda]_\pm. \tag{3.3.23}$$

Note that correlation functions of Q closed operators do not depend on the representative of the class, i.e.

$$\begin{aligned}
\langle \phi_1 \cdots (\phi_k + \{Q, \Lambda\}) \cdots \phi_n \rangle &= \langle \phi_1 \cdots \phi_n \rangle \pm \langle 0 | \phi_2 \cdots \phi_{k-1} \Lambda \phi_{k+1} \cdots \phi_n Q | 0 \rangle \\
&\quad \mp \langle 0 | Q \phi_1 \cdots \phi_{k-1} \Lambda \phi_{k+1} \cdots \phi_n | 0 \rangle \\
&= \langle \phi_1 \cdots \phi_n \rangle.
\end{aligned} \tag{3.3.24}$$

In the last step we have used that the vacuum is annihilated by Q. Another important property of these correlation functions is that they are independent on the position of insertions of the twisted chiral operators and chiral operators. To see this one uses the algebra (3.2.22) and the identity $[\{A,B\},C] = \{[A,C],B\} + \{A,[B,C]\}$ to show

$$\frac{i}{2}\left(\frac{\partial}{\partial x^0} + \frac{\partial}{\partial x^1}\right)\phi = [(H+P),\phi] = [\{Q_+,\overline{Q}_+\},\phi]$$
$$= \cdots = \{Q_B,[Q_+,\phi]\} \qquad (3.3.25)$$

$$\frac{i}{2}\left(\frac{\partial}{\partial x^0} - \frac{\partial}{\partial x^1}\right)\phi = [(H-P),\phi] = [\{Q_-,\overline{Q}_-\},\phi]$$
$$= \cdots = \{Q_B,[Q_-,\phi]\} \qquad (3.3.26)$$

Therefore, we see that the OPE of two (twisted) chiral fields is (twisted) chiral again and position independent. Using a basis ϕ_k of the ring one defines

$$\phi_i \phi_j = C_{ij}^k \phi_k + [Q, \Lambda]_\pm, \qquad (3.3.27)$$

where the structure constants of the ring C_{ij}^k satisfy the usual associativity conditions $C_{jl}^m C_{ik}^l = C_{lk}^m C_{ij}^l$.

These relations can be generalized in the following way. If $\mathcal{O}^{(0)} = \mathcal{O}$ is a Q-closed operator, then one can find a one-form operator $\mathcal{O}^{(1)}$ and a two-form operator $\mathcal{O}^{(2)}$ such that

$$0 = [Q, \mathcal{O}^{(0)}], \qquad (3.3.28)$$
$$d\mathcal{O}^{(0)} = \{Q, \mathcal{O}^{(1)}\}, \qquad (3.3.29)$$
$$d\mathcal{O}^{(1)} = [Q, \mathcal{O}^{(2)}], \qquad (3.3.30)$$
$$d\mathcal{O}^{(2)} = 0. \qquad (3.3.31)$$

In case of the B-twist the one- and two-form operators are given by

$$\text{B-twist:} \left\{ \begin{array}{rl} \mathcal{O}^{(1)} &= idz\,[Q_-,\mathcal{O}] - id\bar{z}\,[Q_+,\mathcal{O}], \\ \mathcal{O}^{(2)} &= dz d\bar{z}\,\{Q_+,[Q_-,\mathcal{O}]\}, \end{array} \right\}, \qquad (3.3.32)$$

and for A-twist they are obtained from the above by the replacement $Q_- \to \overline{Q}_-$. The equations (3.3.28-3.3.31) are called *descent relations*. From Eqs (3.3.29-3.3.30) one deduces that

$$\int_\gamma \mathcal{O}^{(1)} \text{ and } \int_\Sigma \mathcal{O}^{(2)} \qquad (3.3.33)$$

are Q-invariant operators, where γ is a closed one-cycle and Σ is the world-sheet (assumed to have no boundary). These operators will become important when we consider deformations of the twisted topological theories.

3.3.4 Metric (in)dependence and topological string theory

In order to understand the dependence of correlation functions on the worldsheet metric we have to analyze the effect of the energy momentum tensor insertions. The reason is

3.3. TWISTING THE $N = (2,2)$ THEORIES

that classically the energy momentum tensor $T_{\mu\nu} = \frac{1}{\sqrt{h}} \frac{\delta S}{\delta h^{\mu\nu}}$ is the generator of metric variations. Thus the first order variation of the weight factor e^S in the path integral gives the following operator variations

$$\delta_h \langle \mathcal{O} \rangle_g = \langle \mathcal{O} \int_{\Sigma_g} \sqrt{h} d^2\sigma \delta h^{\mu\nu} T_{\mu\nu} \rangle_g, \tag{3.3.34}$$

where g denotes the genus of the worldsheet Riemann surface. In a topological theory $\delta_h \langle \mathcal{O} \rangle_g = 0$ as the energy momentum tensor can always be written in the form

$$T_{\mu\nu} = \{Q, G_{\mu\nu}\}. \tag{3.3.35}$$

This identity is a refinement of the equations (3.3.25)-(3.3.26). In ordinary bosonic string theory the same relation holds with $G_{\mu\nu}$ replaced by the antighost field $b_{\mu\nu}$.

Thus far we have been dealing with topological field theories. In a topological string theory, which will be our main focus in the next chapters, one does not keep the worldsheet metric fixed but rather integrates the metric h of Σ_g over all possible choices of $\mathcal{H}_g$. However, there is a redundancy here as the classical string action is invariant under diffeomorphism and Weyl-transformations of the metric $\tilde{h}_{ab}(\tilde{\sigma}) = \exp[2\omega(\sigma)] \frac{\partial \sigma^c}{\partial \tilde{\sigma}_a} \frac{\partial \sigma^d}{\partial \tilde{\sigma}_b} h_{cd}$. These are "gauge" transformations which even prevail at the quantum level in a critical string theory. The standard method of gauge fixing then reduces the moduli space to

$$\mathcal{M}_g = \text{large gauge tranf.} \backslash \mathcal{H}_g / (\text{diff.} \times \text{Weyl})_g. \tag{3.3.36}$$

Here large gauge transformations refer to diffeomorphisms of Σ_g which are not connected to the identity. For more details on the moduli space and the procedure of gauge fixing see the standard reference [6]. The tangent vectors of $\mathcal{M}_g$ are anti-holomorphic one-forms on Σ_g with values in the holomorphic tangent bundle. These "Beltrami differentials" $\mu_{\bar{z}}^z d\bar{z} \frac{\partial}{\partial z} \in H^1(T_\Sigma)$ parameterize independent first order complex structure deformations of Σ and there are $3g-3$ of them. Denoting the complex structure variables of Σ by $m^a, a = 1, \cdots, 3g-3$ we can thus describe a first order deformation of the metric modulo Weyl and diffeomorphisms as

$$\int_\Sigma d^2\sigma \sqrt{h} \delta h^{ab} T_{ab} = \int_\Sigma d^2 z \mu_{\bar{z}}^{(a)z} \delta m^a T_{zz} + \bar{\mu}_z^{a\bar{z}} \delta \bar{m}^a \bar{T}_{\bar{z}\bar{z}} \tag{3.3.37}$$

which inserted into (3.3.34) gives

$$\frac{\partial}{\partial m^a} \langle \mathcal{O} \rangle_g = \langle \mathcal{O} \int_\Sigma d^2 z \mu_{\bar{z}}^{az} T_{zz} \rangle_g =: \langle \mathcal{O} T^a \rangle_g. \tag{3.3.38}$$

But (3.3.35) means that T is exact and therefore in a cohomological theory any correlation function with a T insertion should vanish. This line of thought would make topological string theory completely trivial and metric independent. It turns out that this is not the case as the calculation (3.3.24) which lies behind the above argument fails at one step. This step is the Q invariance of the vacuum. As will be explained in section 3.5.3 the measure on the moduli space of higher genus Riemann surfaces, which is part of the vacuum definition, is not Q closed.

3.3.5 Dependence on the parameters

In this section we want to outline how topological string correlation functions depend on the target space metric which determines the parameters of the model. In supersymmetric nonlinear sigma models there are three classes of parameters. The first class consists of parameters that enter in D-terms, the second is the class of complex parameters that enter in F-terms (and their conjugates), and finally there is a third class of parameters that determine twisted F-terms (and their conjugates). In the following we will analyze all these from the viewpoint of the B-twisted theory and then state the result for the A-twist.

Let us first consider D-term deformations of the theory. Such a variation inserts in the path-integral an operator of the form

$$\int d^4\theta \Delta K = \int d\bar{\theta}^+ d\bar{\theta}^- d\theta_- d\theta^+ \Delta K. \tag{3.3.39}$$

This is proportional to

$$\left\{\overline{Q}_+, \left[\overline{Q}_-, \int d\theta^+ d\theta^- \Delta K|_{\bar{\theta}^\pm=0}\right]\right\} = \left\{Q_B, \left[\overline{Q}_-, \int d\theta^+ d\theta^- \Delta K|_{\bar{\theta}^\pm=0}\right]\right\}, \tag{3.3.40}$$

where in the last step we have used the nilpotency of $\overline{Q}_-$. This shows that the variation of a D-term leads to a Q_B-exact term in the correlation function and thus vanishes.

Variations of twisted F-terms are induced by twisted chiral operators $\Delta \tilde{W}(\check{\phi})$ annihilated by both $\overline{Q}_+$ and Q_-

$$\int \sqrt{h} d^2 x \int d^2 \tilde{\theta} \Delta \tilde{W}(\tilde{\Phi}) \sim \int \sqrt{h} d^2 x \left\{Q_+, \left[\overline{Q}_-, \Delta \tilde{W}(\check{\phi})\right]\right\}. \tag{3.3.41}$$

Using the fact that $\Delta \tilde{W}(\check{\phi})$ is annihilated by $\overline{Q}_+$ we arrive at

$$\begin{aligned}\left\{Q_+, \left[\overline{Q}_-, \Delta \tilde{W}(\check{\phi})\right]\right\} &= \left\{Q_+, \left[\overline{Q}_- + \overline{Q}_+, \Delta \tilde{W}(\check{\phi})\right]\right\} \\ &= -\left\{Q_B, \left[Q_+, \Delta \tilde{W}(\check{\phi})\right]\right\} + \text{total derivative}, \end{aligned} \tag{3.3.42}$$

where the total derivative term arises from the anti-commutation relations of the supercharges. Therefore, we see that the inserted operator is Q-exact and therefore annihilates any topological correlation function. An analogous statement can be proven for anti-twisted chiral parameters.

The only parameters left are anti-chiral and chiral deformations. The first type is again trivial due to

$$\begin{aligned}\int \sqrt{h} d^2 x \int d^2 \bar{\theta} \Delta \overline{W}(\bar{\phi}) &\sim \int \sqrt{h} d^2 x \left\{\overline{Q}_+, \left[\overline{Q}_-, \Delta \overline{W}(\bar{\phi})\right]\right\} \\ &= \left\{\overline{Q}_+ + \overline{Q}_-, \left[\overline{Q}_-, \Delta \overline{W}(\bar{\phi})\right]\right\} \\ &= \left\{Q_B, \left[\overline{Q}_-, \Delta \overline{W}(\bar{\phi})\right]\right\}, \end{aligned} \tag{3.3.43}$$

3.3. TWISTING THE $N = (2,2)$ THEORIES

where nilpotency of $\overline{Q}_-$ has been used. The chiral parameters are the only parameters on which the topological theory can depend. They lead to the following insertion in correlation functions

$$\int \sqrt{h}d^2x \int d^2\theta \Delta W(\Phi) \sim \int \sqrt{h}d^2x \{Q_+, [Q_-, \Delta W(\phi)]\}$$
$$\sim \int \Delta W(\phi)^{(2)}. \qquad (3.3.44)$$

In the language of CFT this is the second descendant of the chiral operator $\Delta W(\phi)$. It corresponds to the CFT operator (3.2.24) describing complex structure deformations of the Sigma model. Similarly one can show that in the A-twisted theory, topological correlations functions depend holomorphically on twisted chiral parameters corresponding to the operator (3.2.26). As was explained in 3.2.1 these operators describe Kähler structure variations of the Sigma model.

3.3.6 The tt^* equations

The ground states of $N = (2,2)$ two dimensional theories, which also represent the states of the topological theories, change when the theory is deformed by the operators introduced in the previous section. They can be viewed as sitting in the fibers of a vector bundle over the space of deformations. This geometric picture is governed by the tt^* equations which we shall present in the following.

The *operator state correspondence* of 2d CFT associates to every operator ϕ in the ring $\mathcal{R}$ of chiral and twisted chiral operators a state $|\phi\rangle$. Now, there is a canonical way to assign to each state obtained from $\mathcal{R}$ a Ramond-Ramond *vacuum state* defined by

$$Q|\alpha\rangle = Q^\dagger|\alpha\rangle = 0. \qquad (3.3.45)$$

States obeying (3.3.45) have zero energy due to

$$\{Q, Q^\dagger\} = H, \qquad (3.3.46)$$

and the space of such states will be denoted by $\mathcal{V}$ from now on. Let us denote the image of a chiral basis $\phi_i \in \mathcal{R}, i = 0, \cdots, r$ in $\mathcal{V}$ by $|i\rangle$. Then by the operator state correspondence we can also define a representation of the structure constants of the ring (3.3.27) on the vacuum states

$$\phi_i|j\rangle = C_{ij}^k|k\rangle. \qquad (3.3.47)$$

Anti-chiral fields $\bar{\phi}_i \in \mathcal{R}^*$ define an *anti-topological* basis $|\bar{i}\rangle$. The two basis found this way, $|i\rangle$ and $|\bar{i}\rangle$, are related by a linear transformation

$$|i\rangle = M_i^{\bar{i}}|\bar{i}\rangle. \qquad (3.3.48)$$

By CPT we also have $|\bar{i}\rangle = M_{\bar{i}}^j|j\rangle$ where $M_{\bar{i}}^j$ are the matrix elements of the conjugated matrix M^*. Thus, we follow $MM^* = 1$. There are two bilinear pairings among these states. The first is the topological bilinear pairing

$$\langle i|j\rangle = \eta_{ij}, \qquad (3.3.49)$$

and the other is the hermitian bilinear pairing called the tt^* metric

$$\langle \bar{i}|j\rangle = g_{\bar{i}j}. \tag{3.3.50}$$

These two pairings are connected by the formula

$$g^{\bar{l}i}\eta_{ij} = M_j^{\bar{l}}. \tag{3.3.51}$$

The attribute *topological* in the case of the pairing η_{ij} means that this pairing does not depend on the representative of the Q cohomology class. This means that the changes $|i\rangle \mapsto |i\rangle + Q|\lambda\rangle$ or $\langle j| \mapsto \langle j| + \langle \lambda|Q$ do nothing to $\langle i|j\rangle$ as $|j\rangle$ and $\langle i|$ are Q closed. For the paring $g_{\bar{i}j}$ this argument does not apply. The changes $\langle \bar{i}| + \langle \lambda|Q^\dagger$ and $|i\rangle \mapsto |i\rangle + Q|\lambda\rangle$ will not lead to the same result as $|j\rangle$ is not $Q^\dagger$ and $\langle \bar{i}|$ not Q closed. However, the invariance will be maintained once we insert the projector e^{-HT} into the correlation function and take the limit $T \to \infty$. This can be derived from the fact the only states of zero energy are R-R ground states and thus any exact state $Q|\lambda\rangle \neq 0$ will be projected out by the operator e^{-HT} in the $T \to \infty$ limit. For more details on this construction see [36] and [37, 38].

Let us now describe the change in the parameters of the sigma model on $\mathcal{V}$. If we denote the parameters by $m \in \mathcal{M}$, it turns out that the space of the ground states varies as a subspace of the Hilbert space as we vary m. The relevant parameters are the ones appearing in the relevant superpotential and its conjugate

$$S = \int_\Sigma d^2z \mathcal{L}_0 + \sum_i t^i \int_\Sigma d^2z \mathcal{O}_i + \sum_{\bar{i}} \bar{t}^{\bar{i}} \int_\Sigma d^2z \bar{\mathcal{O}}_i, \tag{3.3.52}$$

where by $\mathcal{O}_i$ we denote the two-form descendants. Thus we see that the ground states have the structure of a vector bundle $\mathcal{V}$ over the moduli space parametrized by $m = (t, \bar{t})$. Let e_γ be a basis, then we can define a connection

$$A_{\beta\gamma}^\alpha = g^{\alpha\kappa}\langle e_\kappa|\partial_\beta|e_\gamma\rangle. \tag{3.3.53}$$

To see that A is a connection note that under a change of basis states $|e_\gamma\rangle \mapsto |e_\gamma'\rangle = \Lambda_{\gamma\delta}|e_\delta\rangle$ it undergoes a gauge transformation $A \mapsto \Lambda^{-1}A\Lambda + \Lambda^{-1}d\Lambda$. One can show that in a holomorphic basis the mixed indices of the form $A_{ij}^i = g^{i\bar{k}}\langle \bar{k}|\partial_{\bar{i}}|j\rangle = \eta^{ik}\langle k|\partial_{\bar{i}}|j\rangle$ and $A_{kj}^i = \eta^{il}\langle l|\partial_k|\bar{j}\rangle$ vanish. This in turn shows that the vacuum bundle $\mathcal{V}$ is a holomorphic bundle with a connection compatible with it. The covariant constancy of the metric

$$D_k g_{i\bar{j}} = \partial_k g_{i\bar{j}} - (\partial_k\langle i|)|\bar{j}\rangle - \langle i|\partial_{\bar{k}}|\bar{j}\rangle = \partial_k g_{i\bar{j}} - (\partial_k\langle i|)|\bar{j}\rangle \tag{3.3.54}$$

allows us finally to derive the following formulas for A_{km}^j and $A_{\bar{k}\bar{m}}^j$

$$A_{km}^j = g^{i\bar{j}}\partial_k g_{m\bar{j}}, \quad A_{\bar{k}\bar{m}}^{\bar{j}} = g^{m\bar{j}}\partial_{\bar{k}} g_{m\bar{m}}. \tag{3.3.55}$$

Now we are ready to state the tt^* equations which govern the geometry of the vacuum bundle. The first set of equations is

$$[D_i, D_j] = [\overline{D}_{\bar{i}}, \overline{D}_{\bar{j}}] = 0. \tag{3.3.56}$$

These identities follow from $A_{\bar{i}}$ in case of the holomorphic basis and $A_i = 0$ in case of the anti-holomorphic basis. The most important equation is however

$$[D_i, \overline{D}_{\bar{j}}] = \partial_i A_{\bar{j}} - \bar{\partial}_{\bar{j}} A_i = -[C_i, \overline{C}_{\bar{j}}]. \tag{3.3.57}$$

Here C_i is the matrix of structure constants and the above equation relates the curvature of the vacuum bundle with the structure of chiral/anti-chiral rings. It can be proved by path integral methods in conformal field theory (see [36] for a derivation). There are some more identities and we summarize all below in the topological basis

$$\begin{aligned}
{[D_i, \overline{D}_{\bar{j}}]} &= -[C_i, \overline{C}_{\bar{j}}] \\
[D_i, D_j] = [\overline{D}_{\bar{i}}, \overline{D}_{\bar{j}}] &= [D_i, \overline{C}_{\bar{j}}] = [\overline{D}_{\bar{i}}, C_j] = 0 \\
D_i C_j = D_j D_i \quad & \quad \overline{D}_{\bar{i}} \overline{C}_{\bar{j}} = \overline{D}_{\bar{j}} \overline{C}_{\bar{i}}.
\end{aligned} \tag{3.3.58}$$

This allows us to define a flat $[\nabla_i, \nabla_j] = [\nabla, \overline{\nabla}_{\bar{j}}] = [\overline{\nabla}_{\bar{i}}, \overline{\nabla}_{\bar{j}}] = 0$ connection

$$\nabla_i = D_i + \alpha C_i, \quad \overline{\nabla}_{\bar{j}} = \overline{\nabla}_{\bar{j}} = \overline{D}_{\bar{j}} + \alpha^{-1} \overline{C}_{\bar{j}}. \tag{3.3.59}$$

This flat connection is called the *Gauss Manin connection*. Although it is flat the connection can have monodromies which can make the theory very interesting. We will have more to say about this in later chapters.

3.4 The topological A-model

The topological A-model is important for counting holomorphic curves on the target space manifold as the path integral localizes to holomorphic maps from the worldsheet to the target space. In the M-theory interpretation the A-model can be seen as counting $D2-D0$ bound sates as $D2$ branes wrap holomorphic curves of the Calabi-Yau.

3.4.1 A model without worldsheet gravity

The nonlinear sigma model on a Kähler manifold M of dimension n is described, before twisting, by n chiral multiplet fields Φ^i whose lowest components X^i represent the complex coordinates of the map of the worldsheet to the target space

$$X : \Sigma \to M. \tag{3.4.1}$$

Recall that the A-twist changes the spin of the fermions $\Psi_\pm$ and $\bar{\Psi}_\pm$ as indicated in table 3.3.1. Ψ_- and $\bar{\Psi}_+$ become scalars (denoted by χ^i and $\chi^{\bar{i}}$) while Ψ_+ and $\bar{\Psi}_-$ are anti-holomorphic and holomorphic one-forms (denoted by $\rho^i_{\bar{z}}$ and $\rho^{\bar{i}}_z$) respectively. The action then takes the form

$$S = 2t \int d^2z \left(g_{i\bar{j}} \partial_\nu X^i \partial^\nu X^j + i\epsilon^{\mu\nu} b_{i\bar{j}} \partial_\mu X^i \partial_\nu X^{\bar{j}} - ig_{i\bar{j}} \rho^i_{\bar{z}} D_z \chi^{\bar{j}} - \frac{1}{2} R_{i\bar{k}j\bar{l}} \rho^i_{\bar{z}} \chi^{\bar{l}} \rho^{\bar{k}}_z \chi^j \right), \tag{3.4.2}$$

where a term involving the antisymmetric 2-form $b_{ij} \in H_2(M, \mathbb{Z})$ has been added. The supercharges which remain scalar after the twist are $\overline{Q}_+$ and Q_-. Therefore we set $\bar{\epsilon}_+ = \epsilon_- = 0$ in the supersymmetry transformations (3.2.19). This leaves us with the variation of the fields under $\delta = \bar{\epsilon}_- \overline{Q}_+ + \epsilon_+ Q_-$:

$$\begin{aligned}
\delta X^i &= \epsilon_+ \chi^i, & \delta X^{\bar{j}} &= \bar{\epsilon}_- \chi^{\bar{i}} \\
\delta \rho_{\bar{z}}^i &= 2i\bar{\epsilon}_- \partial_{\bar{z}} X^i + \epsilon_+ \Gamma^i_{jk} \rho_{\bar{z}}^j \chi^k, & \delta \chi^{\bar{i}} &= 0 \\
\delta \chi^i &= 0 & \delta \rho^{\bar{i}} &= -2i\bar{\epsilon}_+ \partial_z X^{\bar{i}} + \bar{\epsilon}_- \Gamma^{\bar{i}}_{\bar{j}\bar{k}} \rho_z^{\bar{k}} \chi^{\bar{j}}
\end{aligned} \quad (3.4.3)$$

with $\delta^2 = 0$.

Physical Operators

Here we want to analyze the Q_A-cohomology classes of operators. We will consider operators which are associated to points on the manifold (i.e. operators of type $\mathcal{O}^{(0)}$ and not $\int_\gamma \mathcal{O}^{(1)}$ or $\int_\Sigma \mathcal{O}^{(2)}$). Thus the only operators left are X and χ to which we can associate differential forms on M according to the rule

$$\chi^i \leftrightarrow dx^i, \chi^{\bar{i}} \leftrightarrow d\bar{x}^{\bar{i}}, \quad (3.4.4)$$

leading to

$$\begin{aligned}
\mathcal{O}_\Omega^{(0)} &= \omega_{i_1 i_2 \cdots i_p \bar{j}_1 \bar{j}_2 \cdots \bar{j}_q}(X) \chi^{i_1} \chi^{i_2} \cdots \chi^{i_p} \chi^{\bar{j}_1} \chi^{\bar{j}_2} \cdots \chi^{\bar{j}_q} \\
&\leftrightarrow \\
\Omega &= \omega_{i_1 i_2 \cdots i_p \bar{j}_1 \bar{j}_2 \cdots \bar{j}_q}(x) dx^{i_1} \wedge dx^{i_2} \cdots dx^{i_p} \wedge d\bar{x}^{\bar{j}_1} \wedge d\bar{x}^{\bar{j}_2} \cdots \bar{x}^{\bar{j}_q}.
\end{aligned} \quad (3.4.5)$$

One checks that under this correspondence Q_- and $\overline{Q}_+$ are identified with the exterior derivatives of Dolbeault cohomology ∂ and $\bar{\partial}$. From this it follows that $Q_A = Q_- + \overline{Q}_+$ is identified with the de Rham operator $d = \partial + \bar{\partial}$ and the operator of Q_A is identified with the exterior derivative

$$\{Q_A, \mathcal{O}_\Omega\} = -\mathcal{O}_{d\Omega}. \quad (3.4.6)$$

Thus Q_A cohomology classes are mapped to d-cohomology classes of differential forms

$$\{\text{physical operator}\} \cong H^*_{DR}(M). \quad (3.4.7)$$

Correlation functions and selection rules

Correlation functions of physical operators $\mathcal{O}_i$ are given by

$$\langle \mathcal{O}_1 \cdots \mathcal{O}_s \rangle = \int \mathcal{D}X \mathcal{D}\chi \mathcal{D}\rho \, e^{-S}, \quad (3.4.8)$$

where the path-integral is taken over all possible configurations. These configurations split into different topological sectors classified by the homology class of the map X from Σ go M:

$$\beta = X_*[\Sigma] \in H_2(M, \mathbb{Z}). \quad (3.4.9)$$

3.4. THE TOPOLOGICAL A-MODEL

Let us now turn to the discussion of selection rules for correlators. Looking at table 3.3.1 we see that χ^i has vector R-charge $q_V = -1$ and axial R-charge $q_A = 1$. Because of the splitting of the tangent bundle of M, namely $TM = TM^{(1,0)} \oplus TM^{(0,1)}$ we can associate to $\mathcal{O}_{\Omega_k}$ an element in the Dolbeault cohomology group $H^{(p_k, q_k)}(M)$. Since $U(1)_V$ remains a symmetry even in the quantum theory, all correlation functions must be invariant under this symmetry and we get $q_v = \sum_{k=1}^n p_k - \sum_{k=1}^n q_k = 0$. However this does not hold for the $U(1)_A$ symmetry as this symmetry is anomalous at quantum level. The anomaly is given by

$$\begin{aligned} q_A &= \sum_{k=1}^n p_k + \sum_{k=1}^n q_k \\ &= \#(\chi\text{zero modes}) - \#(\rho\text{zero modes}) = 2(h^0(X^*(\mathcal{T}_M)) - h^1(X^*(\mathcal{T}_M))) \\ &= 2 \int_\Sigma ch(X^*(TM^{(1,0)}))td(T\Sigma) = 2(c_1(\mathcal{T}_M) \cdot \beta + \dim_\mathbb{C} M(1-g)). \end{aligned} \quad (3.4.10)$$

The combination of the two constraints now gives

$$\sum_{k=1}^n q_k = \sum_{k=1}^n p_k = c_1(\mathcal{T}_M) \cdot \beta + \dim_\mathbb{C} M(1-g). \quad (3.4.11)$$

Localization to Q-fixed points

As there is a fermionic symmetry Q under which all inserted operators are invariant, we can use the localization principle outlined in section 3.3.1 to compute the contributions picked up by the path-integral (3.4.8). We only get contributions from the loci where the Q-variation of the fields vanishes. Looking at (3.4.3) we see that first of all we have to set $\chi = 0$ in order for δX to vanish. Then as a second step one sees that the vanishing of $\delta \rho_z^i$ and $\delta \rho_{\bar{z}}^i$ is equivalent to

$$\partial_{\bar{z}} X^i = 0. \quad (3.4.12)$$

The interpretation of this equation is that the path integral localizes on configurations where the map $X : \Sigma \to M$ is holomorphic. The bosonic part of the action is given by

$$\begin{aligned} S_B &= \int_\Sigma g_{i\bar{j}} \left(\partial_z X^i \partial_{\bar{z}} X^{\bar{j}} + \partial_{\bar{z}} X^i \partial_z X^{\bar{j}} \right) \\ &= 2 \int_\Sigma g_{i\bar{j}} \partial_{\bar{z}} X^i \partial_z X^{\bar{j}} + \int_\Sigma X^*(\omega) \geq \int_\Sigma X^*\omega = \omega \cdot \beta, \quad (3.4.13) \end{aligned}$$

where ω is the Kähler form.

One can see that for holomorphic maps (3.4.12) this action reduces to $\int_\Sigma X^*(\omega) = \omega \cdot \beta$ which shows that the path integral is reduced to finite dimensional integrals over an in general infinite series of components of moduli spaces of holomorphic maps labeled by

$$\mathcal{M}_{g,\beta}(M) = \left\{ X : \Sigma \to M \;\middle|\; \begin{array}{l} \text{holomorphic} \\ X_*[\Sigma] = \beta \end{array} \right\}. \quad (3.4.14)$$

Furthermore this shows that correlation functions are only dependent on the complexified Kähler class $\omega_\mathbb{C} = \omega - iB$.

Let us turn to the calculation of the dimension of this moduli space. Assume for simplicity that the number of χ zero modes is positive and the that the number of ρ zero modes is zero. Then one deduces from (3.4.3) that deformations which preserve the holomorphic structure must satisfy

$$\partial_{\bar{z}} \chi^i = 0. \quad (3.4.15)$$

Thus we see that the tangent space of the moduli space $\mathcal{M}_{g,\beta}(M)$ is identified as the space of χ zero modes yielding

$$\dim_\mathbb{C} \mathcal{M}_{g,\beta} = \#\chi\text{-zero modes}. \quad (3.4.16)$$

The path integral (3.4.8) splits into path integrals over the finite dimensional spaces $\mathcal{M}_{g,\beta}(M)$. In the following we want to identify the measure. Consider the evaluation map at $x_i \in \Sigma$

$$\text{ev}_i : \mathcal{M}_{g,\beta}(M) \to M \quad (3.4.17)$$
$$X \mapsto X(x_i). \quad (3.4.18)$$

Then the operator $\mathcal{O}_{\Omega,i}$ inserted at $x_i \in \Sigma$ can be viewed as the pull-back of $\omega_i \in H^*(M)$ by the evaluation map and the correlation function turns out to be given by

$$\langle \mathcal{O}_{\Omega_1} \cdots \mathcal{O}_{\Omega_s} \rangle_\beta = e^{-i(\omega - iB)\cdot\beta} \int_{\mathcal{M}_{g,\beta}(M)} \text{ev}_1^* \Omega_1 \wedge \cdots \wedge \text{ev}_s^* \Omega_s. \quad (3.4.19)$$

Defining cycles D_i in M as Poincare duals of $[\Omega_i]$, Ω_i can be chosen to have delta function support on D_i and the integral can be identified as the number of holomorphic maps of degree β where x_i is mapped into D_i. Looking at formula (3.4.11) we see that for Calabi-Yau manifolds $c_1(\mathcal{T}_M) = 0$ and the genus $g = 0$ sector one can have a non-vanishing coupling $\langle \mathcal{O}_{\Omega_i} \mathcal{O}_{\Omega_j} \mathcal{O}_{\Omega_k} \rangle$, where all Ω_l are $(1,1)$-forms. Furthermore, with β denoting the cohomology class of the image of the worldsheet in M denoted by C we can write $\beta \cdot \omega = 2\pi \sum_{k=1}^{h^{1,1}} t_k d_k$, where $d_k = C \cap D_k$ is the number of intersections of C with D_k or in other words the degree. Let us look at the map with $d_k = 0$ for all k. This maps the sphere with three punctures $\Sigma_{3,0}$ to a point in M. Together with the fact that $\mathcal{O}_{\Omega_k,k}$ maps to D_k this implies that the path integral collapses to the intersection number of $D_i \cap D_j \cap D_k$. Thus the correlation function can be rewritten in terms of $q_k = e^{-2\pi i t_k}$ as

$$C_{ijk}(t) = \langle \mathcal{O}_{\Omega_i} \mathcal{O}_{\Omega_j} \mathcal{O}_{\Omega_k} \rangle = D_i \cap D_j \cap D_k + \sum_{d_i \neq 0} r_{d_i}^{g=0}(D_i, D_j, D_k) \prod_{i=1}^{h^{1,1}} q_i^{d_i}. \quad (3.4.20)$$

This is a deformation of the classical intersection ring and is known as the *quantum cohomology ring* of M. The second term is called the instanton correction and its effect is to smooth the structure functions $C_{ijk}(t)$ at singularities in codimension two in M [39].

3.4. THE TOPOLOGICAL A-MODEL

Let us now turn to deformations of correlation functions. We observe from the tables (3.3.1,3.3.2) that the second descendants $\mathcal{O}^{(2)}_{\Omega_j}$ have trivial $U(1)_V$ and $U(1)_A$ charges. From this it follows that nontrivial derivatives of $C_{ijk}(t)$

$$\frac{\partial}{\partial t^i}\langle \mathcal{O}_{\Omega_l}\mathcal{O}_{\Omega_k}\mathcal{O}_{\Omega_j}\rangle\bigg|_{t^i=0} = \langle \mathcal{O}_{\Omega_j}\mathcal{O}_{\Omega_k}\mathcal{O}_{\Omega_l}\int_\Sigma \mathcal{O}^{(2)}_{\Omega_i}\rangle \tag{3.4.21}$$

do exist according to the selection rules. Using $SL(2,\mathbb{C})$ invariance on S^2 one can fix three points among the $\{i,j,k,l\}$ and integrate over the fourth yielding

$$\partial_i C_{jkl}(t) = \partial_j C_{ikl}(t). \tag{3.4.22}$$

This is an integrability condition which guarantees the existence of a function $F^0(t)$ obeying the property

$$C_{ijk}(t) = \partial_i \partial_j \partial_k F^0. \tag{3.4.23}$$

$F^0(t)$ is the quantum corrected version of (3.1.52).

3.4.2 Coupling to topological gravity

Here we shall concentrate on the case $c_1(\mathcal{T}_M) = 0$, i.e. we will be dealing with Calabi-Yau manifolds. So far we have been ignoring the degrees of freedom of the worldsheet metric in our discussion. However, as explained in 3.3.4 this would make our theory completely trivial and all correlation functions would vanish for $g > 1$. In order to get a feeling for what is going on let us first have a look at the $g = 1$ case with $\Sigma = T^2$ and target space $M = T^2$ as well. By definition there would be no holomorphic maps between Σ and M unless the complex structure parameter of Σ, denoted by τ_Σ, is equal to the one of M, denoted by τ_M. The free energy for genus 1 would then be $\mathcal{F}^{(1)} = -\log(\eta(\tau_M))\delta(\tau_M - \tau_\Sigma)$ due to multicovering contributions. Integrating this over the complex structure of Σ then gives a result which depends in a modular way on the target space parameter τ_M as the unique parameter. Let us now turn to $g > 1$. According to equation (3.4.10) the axial anomaly becomes negative giving no room for correlation functions to stay nontrivial by insertion of operators. Looking at the problem from the mathematics point of view the dimension of the moduli space $\mathcal{M}_{g,n,\beta}(M)$ of maps from a Riemann surface of genus g with n punctures into the class $\beta = [X(\Sigma)]$ of the manifold M is given by

$$\begin{aligned}\text{vdim}_\mathbb{C}\overline{\mathcal{M}}_{g,n,\beta} &= h^0(X^*(\mathcal{T}_M)) - h^1(X^*(\mathcal{T}_M)) + \dim \text{Def}(\Sigma,\underline{p}) - \dim \text{Aut}(\Sigma,\underline{p}) \\ &= c_1(\mathcal{T}_M) \cdot \beta + (\dim_\mathbb{C} M - 3)(1-g) + n.\end{aligned} \tag{3.4.24}$$

This formula demands some explanation. Here $\text{vdim}_\mathbb{C}$ denotes the virtual complex dimension of the stable compactification $\overline{\mathcal{M}}_{g,\beta,n}$ of the moduli space. By stable compactification we mean that all boundary points where the Riemann surface degenerates are added to the moduli space. $h^0(X^*(\mathcal{T}_M))$ measures deformations of the map X and $h^1(X^*(\mathcal{T}_M))$ measures obstructions of the same map. Furthermore $\text{Def}(\Sigma,\underline{p})$ is the space of deformations of Σ with marked points $\underline{p}$ and $\text{Aut}(\Sigma,\underline{p})$ is the relevant automorphism space. In the last step the Riemann Roch formula has been used to compute the two indices.

Thus we see that Calabi-Yau threefolds play a special role in topological string theory. Substituting $c_1(\mathcal{T}_M) = 0$, $\dim_{\mathbb{C}} M = 3$ and $n = 0$ we get $\text{vdim}_{\mathbb{C}} \overline{\mathcal{M}}_{g,\beta,0} = 0$ which means that the integration over moduli space reduces to the contribution of a few single points. Intuitively one can understand this result as follows. The complex dimension of the moduli space of maps for a fixed Riemann surface ($g \geq 2$) in these cases is formally negative: $\dim_{\mathbb{C}} \mathcal{M} = (\dim M)(1-g) = 3(1-g) < 0$. On the other hand the dimension of the moduli space of metrics on a genus $g \geq 2$ surface has dimension $3(g-1)$. So in these cases the integration over the complex structure of the Riemann surface exactly cancels the positive violation of axial charge.

3.4.3 Target space perspective

According to our discussion in chapter 2 type II string compactifications on Calabi-Yau threefolds lead to $N = 2$ supergravity in four dimensions. Now, it can be shown that certain superpotential terms in the four-dimensional effective theory are captured by topological string amplitudes. In other words for these terms the string integral over the space of all maps from Riemann surfaces to the target space manifold reduces to the A-model topological string result. In particular it can be shown [40, 41] that in the four-dimensional action there is a term generated which looks like

$$\int d^4x d^4\theta \mathcal{W}^{2g} F^g(t_i) = \int d^4x F^g(t_i) R_+^2 F_+^{2g-2} + \cdots, \quad (3.4.25)$$

where $F^g(t_i)$ denotes the genus g topological amplitude depending on the vector moduli t_i. Furthermore, F_+ and R_+ denote the self-dual parts of the graviphoton field strength and the Riemann tensor respectively.

The Gopakumar-Vafa invariants

The Gopakumar-Vafa invariants are integer numbers capturing the BPS content of topological string amplitudes. They arise in an expansion of the one-loop quantum correction to the graviton graviphoton couplings (3.4.25) where BPS states are running in the loop. The way they are encoded in the topological string amplitudes can thus solely be extracted from a supergravity or field theory calculation first performed in [42, 43]. Let us see in some detail how this works.

In passing from the full string theory to its field theory limit a key point lies in the fact that one has to integrate out nonperturbative states of the full string Hilbert space. In our case - namely type II A theory on Calabi-Yau manifolds - these are solitonic states which arise by wrapping $D2$ branes over Calabi-Yau two-cycles. In order for these states to be BPS the two cycles in question must be holomorphic. Compactification of type IIA on a Calabi-Yau M can be equivalently viewed as compactifying M-theory on $M \times S^1$. This way one gets a $U(1)$ gauge field for each 2-cycle of M, obtained by dimensional reduction

3.4. THE TOPOLOGICAL A-MODEL

of the M-theory 3-form C on the 2-cycle, i.e. via the ansatz $C_{\mu\alpha\beta} = A_\mu \omega_{\alpha\beta}$, where $\omega_{\alpha\beta}$ is the harmonic 2-form dual to the 2-cycle. This way a D2-brane wrapped on a 2-cycle (equivalent to a $M2$ brane wrapped on the same cycle) gives a particle charged under the corresponding $U(1)$. Therefore we see that the charges in the theory are classified by the second homology of M, namely $Q \in H_2(M, \mathbb{Z})$. The mass of such a particle is given by

$$m = \frac{1}{\lambda} \int_Q k = \frac{1}{\lambda} t_Q, \qquad (3.4.26)$$

where λ is the string coupling constant and where k denotes the Kähler form on M. Moreover the particle is charged under the graviphoton field with its charge being equal to its mass. Now the effective action corresponding to integrating out these particles at one loop can be computed in a very similar manner to the Schwinger computation where charged particles are integrated out in the presence of the electromagnetic field (see for example [44]). In our case - as we will soon see - the charged particle has to transform in a non-trivial representation R of the four-dimensional Lorentz group. At the Lie-algebra level we have $SO(4) = SU(2)_L \times SU(2)_R$ and as F_+ only couples to the left-handed representation, only the $SU(2)_L$ content of the representation R will enter our formulas. Thus Schwinger's calculation carried out with our setup gives

$$S = \log \det(\Delta + m^2 + \sigma_L F_+) = \int_\epsilon^\infty \frac{ds}{s} \frac{\text{Tr}(-1)^F e^{-sm^2} e^{-2se\sigma_L F_+}}{(2\sinh(seF_+/2))^2}. \qquad (3.4.27)$$

This formula demands some explanation. First of all e is the charge of the particle and σ_L denotes the Cartan element of $SU(2)_L$. Second, we have to take into account that fermions and bosons have opposite powers of determinant which leads to the insertion of $(-1)^F$. Note that we are dealing with a supersymmetric theory and the particles running in the loop are BPS and thus half-hypermultiplets transforming in a representation

$$[(1/2, 0) + 2(0, 0)] \otimes \mathcal{R}, \qquad (3.4.28)$$

where $\mathcal{R}$ is some representation of $SO(4)$. The other difference to the ordinary Schwinger calculation is that we have a further $\int R_+^2$ insertion. Luckily it turns out that the extra R_+^2 insertion absorbs the representation $[(1/2, 0) + 2(0, 0)]$ and leaves us only with $\mathcal{R}$. The same would happen if we were dealing with a vector multiplet running in the loop - the only difference being that we would get a minus sign due to the presence of an extra fermion. Our results reduce to the non-supersymmetric Schwinger calculation and we get formula (3.4.27) with the substitutions $e = m$, $m = t_Q/\lambda$, and $F_+\lambda \mapsto \lambda$, i.e.

$$S = \int F(t, \lambda) R_+^2, \text{ where } F(t, \lambda) = \int_\epsilon^\infty \frac{ds}{s} \frac{\text{Tr}(-1)^F e^{-st_Q} e^{-2s\sigma_L \lambda}}{(2\sinh(s\lambda/2))^2}. \qquad (3.4.29)$$

In order to extract the final result let us switch to the M-theory picture. Upon compactification on the Calabi-Yau M M-theory reduces to a five dimensional effective theory and M2 branes wrapping two-cycles of the Calabi-Yau will correspond to particles transforming under the rotation group $SO(4) \sim SU(2)_L \times SU(2)_R$. In this picture D2 branes arise from M2 branes through dimensional reduction on the extra S^1. As each such M2-brane can have an additional momentum of n units around the extra circle for each M2 brane we get infinitely many D2-branes labeled by n. The masses of these branes are then proportional to $|t_Q + 2\pi i n|$.

Next, we introduce numbers $n^Q_{(j_L, j_R)}$ which count the number of BPS M2-branes in the class Q giving rise to particles transforming in the representation

$$[(1/2, 0) + 2(0, 0)] \otimes (j_L, j_R). \tag{3.4.30}$$

From this we can deduce the numbers

$$n^Q_{j_L} = \sum_{j_R} (-1)^{(2j_R)} (2j_R + 1) n^Q_{(j_L, j_R)}, \tag{3.4.31}$$

which are invariant under smooth deformations of the theory (this will be explained in more detail in the next subsection).

For the $SU(2)_L$ representation we will choose the basis

$$I_r = I_1^{\otimes r} = [(1/2) + 2(0)]^{\otimes r}, \tag{3.4.32}$$

whose significance will again become clear when we switch to the geometric point of view in the next subsection. Thus we arrive at the following redefinition of the numbers $n^Q_{n_L}$

$$\sum_{r=0}^{\infty} n^Q_r I_r = \sum_{j_L} n^Q_{j_L} [j_L], \tag{3.4.33}$$

where the integer numbers n^Q_r are implicitly defined. Using the identities

$$\begin{aligned} \text{Tr}_{I_1}(-1)^F e^{-2s\sigma_L \lambda} &= 2 - e^{-2s} - e^{2s} = [2i \sinh(s\lambda/2)]^2, \\ \Rightarrow \text{Tr}_{I_r}(-1)^F e^{-2s\sigma_L \lambda} &= \left[\text{Tr}_{I_1}(-1)^F e^{-s\sigma_L \lambda}\right]^r = [2i \sinh(s\lambda/2)]^{2r}, \end{aligned} \tag{3.4.34}$$

and the expression

$$\sum_n \exp(-2\pi i n s) = \sum_m \delta(s - m), \tag{3.4.35}$$

we see that for each wrapped D2-brane in the charge class Q and in the representation I_r the contribution to $F(t)$ is given by

$$\sum_n \int \frac{ds}{s} e^{-s(t_Q + 2\pi i n)} [2i \sinh(s\lambda/2)]^{2r-2} = \sum_{m \geq 0} \frac{1}{m} e^{-mt_Q} [2i \sinh(m\lambda/2)]^{2r-2}. \tag{3.4.36}$$

Adding to the above result the contribution of unwrapped D2-branes living in the representation I_0 and the classical terms we obtain as final result

$$\begin{aligned} F(t, \lambda) &= \frac{1}{\lambda^2} \left[\frac{1}{6} \kappa_{ijk} t^i t^j t^k + P_2(t) \right] + \frac{-1}{24} c_2^i t_i + \text{const.} \tag{3.4.37} \\ &+ \sum_{g > 1} \frac{-\chi B_g B_{g-1}}{4g(2g-2)(2g-2)!} \lambda^{2g-2} + \sum_{m, Q, r} \frac{e^{-mt_Q}}{m [2 \sin(m\lambda/2)]^{2-2r}} n^Q_r \tag{3.4.38} \end{aligned}$$

where B_g is the gth Bernoulli number and the sum on the RHS is over all $m > 0$, $r \geq 0$, $Q \in H_2(M, \mathbb{Z})$ and the n^Q_r are all integers and are known as Gopakumar-Vafa invariants.

3.4. THE TOPOLOGICAL A-MODEL

Geometric interpretation

As noted in the previous section the n_r^Q capture the $SU(2)_L$ content of the number of wrapped BPS D2-branes with charge $Q \in H_2(M, \mathbb{Z})$ in a particular basis for the $SU(2)_L$ representation ring. In this section we want to give a geometric interpretation of this following naturally from the worldvolume theory of the D2 branes. The bosonic sector of the worldvolume theory of a single D2-brane consists of 7 scalars and one $U(1)$ gauge field. Among the 7 scalars four of them describe deformations of the D-brane within the Calabi-Yau and the other three are movements in the uncompactified directions. Let us from now on assume that the D2-brane is wrapping a genus g Riemann surface Σ. Then the quantum of flux of the $U(1)$ field strength on the Riemann surface is the number of D0-branes bound to the D2 brane. Focusing on the movement of the D2-brane within the Calabi-Yau the worldvolume theory is a supersymmetric sigma model with target space $\hat{\mathcal{M}}$, which is the moduli space of deformations of Σ together with a choice of a flat $U(1)$ connection on Σ. Denoting by $\mathcal{M}$ the moduli space of deformations we thus see that we have the following fibration structure

$$\hat{\mathcal{M}} \to \mathcal{M}, \tag{3.4.39}$$

where the fiber is generically T^{2g}, i.e. the Jacobian of Σ. As usual in supersymmetric sigma models the number of bound states (ground states) in this formulation is in exact correspondence with the cohomology of $\hat{\mathcal{M}}$. Our aim is now to give the $SU(2)_L$ content of the cohomology of this moduli space. Note that Kähler manifolds admit an $SU(2)$ action of their cohomology (denoted by Lefschetz action), where the $SU(2)$ raising operator J_+ corresponds to wedging with the Kähler class, J_- corresponds to its adjoint and J_3 acting on $H^{p,q}$ has eigenvalue given by

$$(p + q - \dim_{\mathbb{C}}\hat{\mathcal{M}})/2. \tag{3.4.40}$$

In the case of the Kähler manifold $\hat{\mathcal{M}}$ the Lefschetz action on the base $\mathcal{M}$ is identified with our previous $SU(2)_R$. This can be seen most easily in the M-theory picture. The scalars parameterizing the deformations of Σ in the threefold are paired with fermions who have quantum numbers $(0, \frac{1}{2})$ under the spacetime rotation group $SO(4) \cong SU(2)_L \otimes SU(2)_R$. Furthermore, quantizing the fermions which come in pairs with one-forms of Σ gives rise to the representation

$$\left[(\frac{1}{2},0) + 2(0,0)\right]^{g+1}, \tag{3.4.41}$$

which transforms under $SU(2)_L$ of the four dimensional rotation group. Note that this is exactly the representation of a BPS-hypermultiplet tensored with I_g. As $I_g = \left[(\frac{1}{2},0) + 2(0,0)\right]^g$ is the $SU(2)$ content of $T^{2g} = \text{Jac}(\Sigma)$ we thus see that the $SU(2)_L$ is identified with the fiber Lefschetz action.

Let us now outline how one can use this geometric picture to compute the Gopakumar-Vafa invariants n_r^Q. The first point will be to relate $H^*(\Sigma)$ to $H^*(\text{Jac}(\Sigma))$. Denoting the

Kähler class of the fiber of $\hat{\mathcal{M}}$ by θ one can see easily the following identity

$$\theta^{g-1}\mathrm{Jac}(\Sigma) = \binom{1}{2} \oplus 2g(0), \tag{3.4.42}$$

which combined with $H^*(\Sigma) = \binom{1}{2} \oplus 2g(0)$ gives

$$H^*(\Sigma) = \theta^{g-1}\mathrm{Jac}(\Sigma). \tag{3.4.43}$$

Note that summand (a) on the right hand side of (3.4.42) denotes a vector space with weights equal to the weights of the (a) representation shifted by a certain amount. In order to generalize (3.4.43) to actions of θ^r on $\mathrm{Jac}(\Sigma)$ it turns out that one has to look at the pth symmetric product of Σ, namely $\mathrm{Sym}^p(\Sigma)$. The symmetric product of Σ consists of p unordered points of Σ. As an example let us look at the second symmetric product. Then we have $H^*(\mathrm{Sym}^2(\Sigma)) \cong \mathrm{Sym}^2 H^*(\Sigma)$, giving

$$H^*(\mathrm{Sym}^2\Sigma) = (1) \oplus 2g\binom{1}{2} \oplus \binom{2g}{2}(0). \tag{3.4.44}$$

Again an easy calculation using induction shows

$$\theta^{g-2}I_g = (1) \oplus 2g\binom{1}{2} \oplus (2g^2 - g - 1)(0), \tag{3.4.45}$$

which combined with $H^*(\mathrm{Sym}^0\Sigma) = (0)$ and (3.4.44) yields

$$H^*(\mathrm{Sym}^2\Sigma) = \theta^{g-2}I_g \oplus H^*(\mathrm{Sym}^0\Sigma). \tag{3.4.46}$$

This equation can be generalized and one arrives at the following result

$$H^*(\mathrm{Sym}^p\Sigma) = \theta^{g-p}H^*(\mathrm{Jac}(\Sigma)) \oplus H^*(\mathrm{Sym}^{p-2}\Sigma). \tag{3.4.47}$$

The symmetric product is well defined in this case only for smooth curves. For singular curves we have to use the Hilbert scheme $\mathrm{Hilb}^p(\Sigma)$ which coincides with the former for smooth curves. An element of $\mathrm{Hilb}^p(\Sigma)$ consists of the curve Σ together with the choice of p points on it. Letting Σ vary in a family $\mathcal{M}$ we arrive at the *relative Hilbert scheme*

$$\mathcal{C}^{(p)} = \{(\Sigma, Z)|\Sigma \in \mathcal{M}, Z \in \mathrm{Hilb}^p(\Sigma)\}. \tag{3.4.48}$$

As $\mathrm{Sym}^p(\Sigma)$ varies as the fibers of the $\mathcal{C}^{(p)}$ over $\mathcal{M}$ and $\mathrm{Jac}(\Sigma)$ varies as the fibers of the family $\hat{\mathcal{M}}$ over $\mathcal{M}$ equation (3.4.47) generalizes to

$$H^*(\mathcal{C}^{(p)}) = \theta^{(g-p)}H^*(\hat{\mathcal{M}}) \oplus H^*(\mathcal{C}^{(p-2)}). \tag{3.4.49}$$

Now, as the definition of the $n^r_{[\Sigma]}$ reads

$$H^*(\hat{\mathcal{M}}) = \sum n^r_{[\Sigma]}I_r, \tag{3.4.50}$$

we can use equation (3.4.49) to compute the $n^r_{[\Sigma]}$ in explicit cases. Consider the case $p = 0$, then equation (3.4.49) simplifies to

$$H^*(\mathcal{M}) = \theta^g H^*(\hat{\mathcal{M}}) \tag{3.4.51}$$

as $\mathcal{C}^{(0)} = \mathcal{M}$. Applying $\text{Tr}(-1)^F$ to this we arrive at

$$(-1)^{\dim \mathcal{M}} e(\mathcal{M}) = n^g_{[\Sigma]}, \tag{3.4.52}$$

where $e(\mathcal{M})$ is the Euler characteristic of $\mathcal{M}$. This is a simple and important formula which is a general result in the theory of Gopakumar-Vafa invariants. Next, we turn to the case $p = 1$. Then equation (3.4.49) reads

$$\begin{aligned} H^*(\mathcal{C}) &= \theta^{g-1} H^*(\hat{\mathcal{M}}) \\ &= \theta^{g-1}(n^g_{[\Sigma]} I_g + n^{g-1}_{[\Sigma]} I_{g-1} + \cdots) \\ &= n^{g-1}_{[\Sigma]}(0) \oplus n^g_{[\Sigma]}\left(\left(\frac{1}{2}\right) \oplus 2g(0)\right). \end{aligned} \tag{3.4.53}$$

Again applying $\text{Tr}(-1)^F$ to this, we get

$$(-1)^{\dim \mathcal{M}+1} e(\mathcal{C}) = (2g-2)n^g_{[\mathcal{C}]} + n^{g-1}_{[\mathcal{C}]}, \tag{3.4.54}$$

where it was used that $\dim \mathcal{C} = \dim \mathcal{M} + 1$. We will not delve more into these calculations at this point but will refer the reader to the beautiful expositions [45] and [19].

3.4.4 Interpretation around the Conifold singularity

Calabi-Yau manifolds admit so called *Conifold singularities*, see section 3.1.2, where in the A model the local geometry near the singularity is given by

$$\begin{array}{c} \mathcal{O}(-1) \oplus \mathcal{O}(-1) \\ \downarrow \\ \mathbb{P}^1 \end{array}. \tag{3.4.55}$$

Near a conifold singularity, the metric of the moduli space is degenerate and the prepotential takes the following form

$$F_c^0 \sim t_c^2 \log t_c + \mathcal{O}(t_c^0), \tag{3.4.56}$$

where t_c is the Kähler parameter of the shrinking cycle. Strominger [20] argued that the occurrence of such a singularity in the four dimensional effective action can be traced back to integrating out light Ramond-Ramond black holes with mass t_c. A further argument by Vafa [54] shows that the genus one free energy is singular as well for $t_c \to 0$ due to integrating out the same massless hypermultiplet, namely we have

$$F_c^1 = -\frac{1}{12} \log t_c + \mathcal{O}(t_c). \tag{3.4.57}$$

For higher genus calculations one can use (3.4.55) to show that for any class $Q = d[\mathbb{P}^1]$ the moduli space $\hat{\mathcal{M}}$ of deformations is just a point. Consider first the case $d = 1$. As the local geometry (3.4.55) has no complex structure deformations, the base $\mathcal{M}$ is trivial. Also, there are no flat connections on the $\mathbb{P}^1$ which shows that the fibre is trivial as well. Now look at the space of stable rank d bundles on $\mathbb{P}^1$. This set is also empty, the physical interpretation being that there are no $D2$-brane bound states on $\mathbb{P}^1$. We therefore see that there is only one hypermultiplet getting massless at the conifold and is sitting in the representation $[(1/2, 0) + 2(0, 0)]$. Denoting the size of the $\mathbb{P}^1$, i.e. the mass of the hypermultiplet by t_c and following the argumentation in the section about the Gopakumar-Vafa-invariants we arrive at the one loop integral

$$F(\lambda, t_c) = \int_\epsilon^\infty \frac{ds}{s} \frac{\exp(-st_c)}{4\sin^2(s\lambda/2)} + \mathcal{O}(t_c^0) = \sum_{g=2}^\infty \left(\frac{\lambda}{t_c}\right)^{2g-2} \frac{(-1)^{g-1} B_{2g}}{2g(2g-2)} + \mathcal{O}(t_c^0). \quad (3.4.58)$$

This bears the following interpretation. Expanding the topological free energies around Conifold-singularities in the Calabi-Yau moduli space we expect the above gap structure. This phenomenon was first observed in [51] and we will make extensive use of it in this thesis.

3.5 The topological B-model

The topological B-model is much simpler than the A-model as the relevant path integral configurations turn out to be constant maps to the Calabi-Yau. However, the B-model is still very powerful due to mirror symmetry which can be used to translate its correlation functions to A-model amplitudes. Note that the B-model is only consistent for Kähler manifolds with vanishing first Chern class, as the axial $U(1)_A$ which is used for twisting is anomalous with an anomaly proportional to $\int_\Sigma X^*(c_1(T_M))$.

3.5.1 B-model without worldsheet gravity

In the case of the B-model the scalar BRST operator is given by

$$Q_B = \overline{Q}_- + \overline{Q}_+. \quad (3.5.1)$$

Once we go over to new convenient variables

$$\eta^{\bar{i}} := -(\Psi_-^{\bar{i}} + \Psi_+^{\bar{i}}), \quad \theta_j := g_{j\bar{i}}(\Psi_+^{\bar{i}} - \Psi_-^{\bar{i}}), \quad (3.5.2)$$

the action of the Q_B transformation simplifies to

$$\begin{aligned}&\delta X^i = 0,\ \delta\theta_i = 0,\\&\delta X^{\bar{i}} = \bar{\epsilon}\eta^{\bar{i}},\ \delta\eta^{\bar{i}} = 0,\\&\delta\rho_\mu^i = \pm 2i\bar{\epsilon}\partial_\mu X^i.\end{aligned} \quad (3.5.3)$$

Physical operators

The space of physical operators is constructed from X^i, $X^{\bar{i}}$, $\eta^{\bar{i}}$ and θ_i. As the fields $\eta^{\bar{i}}$ and θ_i are connected through the metric $g^{\bar{i}j}$ one immediately sees the correspondence

$$\eta^{\bar{i}} \longleftrightarrow dx^{\bar{i}} \tag{3.5.4}$$

$$\theta_i \longleftrightarrow \frac{\partial}{\partial x^i}. \tag{3.5.5}$$

Therefore, a general operator which is a string in the $\eta^{\bar{i}}$ and θ_i multiplied with an X^i, $X^{\bar{i}}$ dependent field corresponds to

$$\omega^{j_1\cdots j_q}_{\bar{i}_1\cdots\bar{i}_p}\eta^{\bar{i}_1}\cdots\eta^{\bar{i}_p}\theta_{j_1}\cdots\theta_{j_q} \longleftrightarrow \omega^{j_1\cdots j_q}_{\bar{i}_1\cdots\bar{i}_p}d\bar{z}^{\bar{i}_1}\wedge\cdots d\bar{z}^{\bar{i}_p}\frac{\partial}{\partial z^{j_1}}\wedge\cdots\frac{\partial}{\partial z^{j_q}}. \tag{3.5.6}$$

This is an anti-holomorphic p-form with values in the q-th exterior power of the holomorphic tangent bundle $\mathcal{T}_M$ - an element of $\Omega^{0,p}(M,\wedge^q\mathcal{T}_M)$. One can show easily that the operator Q_B is identified with the Dolbeault operator $\bar{\partial}$. This way the Q_B-cohomology is identified as the Dolbeault cohomology groups

$$H^*_{Q_B} = \frac{\mathrm{Ker}Q_B}{\mathrm{Im}Q_B} = \bigoplus_{p,q=0}^n H^{0,p}(M,\wedge^p\mathcal{T}_M). \tag{3.5.7}$$

Correlation functions

In this section we want to obtain selection rules for physical correlation functions and relate them to topological quantities. Consider the correlation function

$$\langle\mathcal{O}_1\cdots\mathcal{O}_s\rangle = \int \mathcal{D}X\mathcal{D}\theta\mathcal{D}\eta e^{-S}\mathcal{O}_1\cdots\mathcal{O}_s, \tag{3.5.8}$$

where the $\mathcal{O}_i$ correspond to $\omega_i \in H^{0,p_i}(M,\wedge^{q_i}\mathcal{T}_M)$. From the $U(1)_V$ symmetry condition it follows that this correlator is non-vanishing if $\sum_{i=1}^s p_i = \sum_{i=1}^s q_i$. On the other hand the $U(1)_A$ symmetry has an anomaly after twisting which implies that $\sum_{i=1}^s(p_i + q_i) = 2\dim_\mathbb{C}M(1-g)$ has to hold for the correlator to be nonvanishing. For genus 0 these two requirements give together

$$\sum_{i=1}^s p_i = \sum_{i=1}^s q_i = \dim_\mathbb{C}M, \tag{3.5.9}$$

whereas for $g = 1$ the correct condition is $\sum_i p_i = \sum_i q_i = 0$.

As a next step we shall evaluate the correlation function (3.5.8) by the principle of localization. Looking at equations (3.5.5) we see that a Q-fixed point obeys

$$\partial_\mu X^i = 0, \tag{3.5.10}$$

which identifies it as a constant map. Thus, as a consequence, the path integral is taken over M as the space of constant maps is M itself. All we need now is to find a canonical measure on M. Note that the operators are not ordinary differential forms but rather $(0,p)$-forms with values in the $\wedge^q \mathcal{T}_M$. Furthermore, for Calabi-Yau threefolds we have $p = q = 3$ and thus the holomorphic three form Ω can be used to "absorb" the tangent indices. This works as follows, define a $(3,3)$-form via

$$\omega \mapsto \langle \omega, \Omega \rangle := \omega^{i_1 \cdots i_n}_{\bar{j}_1 \cdots \bar{j}_n} d\bar{z}^{\bar{j}_1} \wedge \cdots \wedge d\bar{z}^{\bar{j}_n} \Omega_{i_1 \cdots i_n} \wedge \Omega. \tag{3.5.11}$$

For genus 0 this translates to

$$\langle \mathcal{O}_1 \mathcal{O}_2 \mathcal{O}_3 \rangle = \int_M \mu_1^i \wedge \mu_2^j \wedge \mu_3^k \Omega_{ijk} \wedge \Omega, \tag{3.5.12}$$

where $\mu_1, \mu_2, \mu_3 \in H^1(M, \mathcal{T}_M)$ are the Beltrami differentials. Note that this result for the three point function is precisely the third-order derivative of the prepotential of the complex structure moduli space (see section 3.1.2)

$$\partial_1 \partial_2 \partial_3 \mathcal{F}. \tag{3.5.13}$$

3.5.2 Picard-Fuchs equations

The complex structure deformation space for Calabi-Yau manifolds is unobstructed and one can turn on a final deformation starting from a given complex structure [55, 56]. Thus Calabi-Yau manifolds come in families which are smoothly connected. Our aim in this section is to find a parameterization of the holomorphic three-form and its periods in terms of affine coordinates on the deformation space. From the viewpoint of the B-model these can be interpreted as follows. Chiral primary fields of $U(1)$ charge q correspond to elements of $H^{3-q,q}(M)$ which are subspaces of the ordinary de Rham cohomology group $H^3(M, \mathbb{C})$. However, the de Rham cohomology group depends only on the topology of M and is thus independent of the complex structure deformations t^i. On the other hand the charge subspaces $H^{3-q,q}(M)$ do rotate within $H^3(M, \mathbb{C})$ as one moves the t^i's. Thus they form a bundle over the moduli space which in fact can be identified with the bundle $\mathcal{V}$ introduced in section 3.3.6. Fields with charge $(0,0)$ can be identified with the holomorphic 3-form and charge $(1,1)$ fields correspond to $(2,1)$ cohomology classes according to section 3.5.1. Furthermore, the Gauss-Manin connection of section 3.3.6 tells us how the space $\mathcal{V}$ moves within the Hilbert space. Geometrically this is captured by the so called *Picard-Fuchs equations* which are ordinary differential equations annihilating the periods of Ω.

Having described the basic setup we want to specialize to the case where the complex structure moduli space is a compact Riemann surface C where in our exposition we will follow the references [57, 58]. Assuming that the Calabi-Yau manifold M_z over a point $z \in C$ is n-dimensional we choose topological n-cycles $\Gamma_0, \cdots, \Gamma_{r-1}$ which give a basis for the nth homology of one particular M_0. Furthermore we choose a holomorphic n-form Ω on M_0. The existence of the Picard-Fuchs equation can be deduced by the following argument. As closed cycles $\Gamma_i(z)$ can always be chosen to be covariantly constant sections of the homology bundle differentiations with respect to z can be passed over the

3.5. THE TOPOLOGICAL B-MODEL

integral $\int_{\Gamma_i(z)}$ and thus only act on the holomorphic three-form Ω. On the other hand, one knows from the analysis in section (3.1.2) that derivatives of Ω always stay in the finite dimensional subspace $F^0 = H^{3,0} \oplus H^{2,1} \oplus H^{1,2} \oplus H^{0,3}$. Therefore, we see that the vectors

$$\Pi_j(z) := \left[\frac{d^j}{dz^j} \int_{\Gamma_0(z)} \Omega(z), \cdots, \frac{d^j}{dz^j} \int_{\Gamma_{r-1}(z)} \Omega(z) \right] \in \mathbb{C}^r, \tag{3.5.14}$$

where $\Pi_0(z)$ is the *period vector*, remain in a maximally r-dimensional space. Furthermore, for generic values of the parameter z, the dimensions

$$d_j(z) := \dim(\operatorname{span}\{\Pi_0(z), \cdots, \Pi_j(z)\}) \tag{3.5.15}$$

must be constant. Thus we conclude that there will be a smallest s such that

$$\Pi_s(z) \in \operatorname{span}\{\Pi_0, \cdots, \Pi_{s-1}\}. \tag{3.5.16}$$

The last equation can be rewritten as the Picard-Fuchs equation, satisfied by all periods of $\Omega(z)$, namely

$$\Pi_s(z) = -\sum_{j=0}^{s-1} C_j(z) \Pi_j(z) \Leftrightarrow \frac{d^s f}{dz^s} + \sum_{j=0}^{s-1} C_j(z) \frac{d^j f}{dz^j} = 0, \tag{3.5.17}$$

where f is an arbitrary period.

Note that the coefficients $C_j(z)$ may acquire singularities at special values of z. However, it turns out that these singularities are of *regular* type such that multiplication of the Picard-Fuchs operator by z^s turns it into the form

$$(z\frac{d}{dz})^s + \sum_{j=0}^{s-1} B_j(z) (z\frac{d}{dz})^j, \tag{3.5.18}$$

where the redefined coefficients $B_j(z)$ are holomorphic functions of z. This equation can be transformed to the matrix equation

$$z\frac{d}{dz} w(z) = A(z) w(z), \tag{3.5.19}$$

where

$$A(z) = \begin{bmatrix} 0 & 1 & & & \\ & 0 & 1 & & \\ & & \ddots & \ddots & \\ & & & 0 & 1 \\ -B_0(z) & -B_1(z) & \cdots & \cdots & -B_{s-1}(z) \end{bmatrix}, \tag{3.5.20}$$

and

$$w(z) = \begin{bmatrix} f(z) \\ z\frac{d}{dz} f(z) \\ \vdots \\ (z\frac{d}{dz})^{s-1} f(z) \end{bmatrix}. \tag{3.5.21}$$

The solutions to equation (3.5.19) can be written in terms of a constant $s \times s$ matrix R and a single valued $s \times s$ matrix $S(z)$ which consists of functions of z, regular around $z = 0$, as follows

$$\Phi(z) = S(z) \cdot z^R. \qquad (3.5.22)$$

Φ is called the *fundamental matrix* for the system. Furthermore, we have the following expansion

$$z^R := e^{(\log z)R} = I + (\log z)R + \frac{(\log z)^2}{2!}R^2 + \cdots, \qquad (3.5.23)$$

defining a multiple-valued matrix function of z. The matrix $e^{2\pi i R}$ is called the *monodromy matrix* of the system as it gives the local monodromy on the solutions around $z = 0$. For applications of Mirror Symmetry we will particularly be interested in points of *maximal unipotent monodromy*. This is, by definition, a point in moduli space where $(e^{2\pi R} - I)^{m+1}$ is a unipotent matrix, such that $(e^{2\pi R} - I)^m \neq 0, (e^{2\pi i R} - I)^{m+1} = 0$ for $m + 1$ being the order of the Picard-Fuchs-operator.

Generally, one has nontrivial monodromy transformations about the boundary divisors in moduli space. That is, if we consider a closed loop $\gamma \in H_1(C)$ around a boundary point $z_0 \in \overline{C}$ (here $\overline{C}$ is the closure of C), then the period vector transforms as

$$\Pi(z) = M(\gamma)\Pi(z), \quad M(\gamma) \in Sp(h_3, \mathbb{Z}). \qquad (3.5.24)$$

The group generated by transport around all closed loops is denoted by the *monodromy group* $\Gamma \subseteq Sp(h_3, \mathbb{Z})$. In addition, since the monodromy group must be a representation of the fundamental group of the moduli space $\overline{C}$ with boundary points removed, we have the fundamental property

$$\Pi_i M(\gamma_i) = \mathbf{1}_{h_3} \qquad (3.5.25)$$

for loops γ_i which go around all singular loci z_i.

We will not delve more into the theory of Picard-Fuchs operators at this point as the methods for obtaining such operators and solving the relevant equations can be quite specific for distinct Calabi-Yau geometries. In section 4.2 we will give some details on Picard-Fuchs equations for one-parameter models, in section 5.1.2 the theory of Picard-Fuchs equations for local Calabi-Yau manifolds is reviewed and section 6.2 contains a generalization for compact hypersurfaces.

3.5.3 Coupling to topological gravity

In order to couple the B-twisted sigma model to topological gravity we have to integrate over the moduli space $\mathcal{M}_g$ introduced in section 3.3.4. The expected complex dimension of this space is $3g - 3$ and therefore the correct measure will be consisting of a real $6g - 6$ volume form. The way this measure is found is completely analogous to the bosonic string case due to the following correspondence. The structure of the twisted $N = 2$ algebra is isomorphic to the one which is obtained from the BRST quantization of the bosonic string. In particular we have

3.5. THE TOPOLOGICAL B-MODEL

$$(G^+, J, T, G^-) \leftrightarrow (Q, J_{\text{ghost}}, T, b), \qquad (3.5.26)$$

where Q is the BRST generator of the bosonic string, b the antighost corresponding to diffeomorphism symmetry on the string worldsheet, and J_{ghost}, T are the ghost number and the energy momentum tensor respectively. Recall that in the case of the bosonic string after having performed the Faddeev-Popov procedure one obtains the following measure factor

$$\int_{\mathcal{M}_g} \langle | \prod_{i=1}^{3g-3} b(\mu_i) |^2 \rangle, \qquad (3.5.27)$$

where the μ_i are "Beltrami" differentials introduced in section 3.3.4 and $b(\mu_i)$ is given by

$$b(\mu) = \int_{\Sigma_g} b_{zz} \mu_{\bar{z}}^z. \qquad (3.5.28)$$

Note, that the number of b insertions in the correlator is just enough to soak up all zero modes of the b ghost which count $6g - 6 = -3\chi(\Sigma_g)$. Analogously, we introduce the following measure into the B-twisted theory

$$\mathcal{F}^{(g)} = \int_{\mathcal{M}_g} \langle | \prod_{i=1}^{3g-3} G^-(\mu_i) |^2 \rangle, \qquad (3.5.29)$$

where by $\mathcal{F}^{(g)}$ we denote the free energy of the topological theory at genus g. Here we use the fact that $G^-, \overline{G^-}$ have $h = 2$ after B-twist to define

$$G^-(\mu) := \int_{\Sigma_g} G^-_{zz} \mu_{\bar{z}}^z. \qquad (3.5.30)$$

In place of the ghost number anomaly this time we have an axial current anomaly (section 3.3.2) which is again equal to $6g - 6 = -3\chi(\Sigma_g)$. This anomaly is canceled by the insertion of $(6g - 6) \times G^-$.

3.5.4 The holomorphic anomaly equations

Let us first explain what we mean here by anomaly in the holomorphic context. Recall that as stated in section 3.3.5 correlation functions of the B-twisted topological theory do not depend on anti-chiral insertions. This on the other hand implies that deformations of the Sigma-model Lagrangian do not depend on anti-chiral parameters and thus that antiholomorphic derivatives of the free energies should vanish. However, as argued in section 3.3.4 this relies crucially on the Q-closeness of the vacuum. In this section we shall see that insertions of anti-chiral fields lead to an integration over the boundary components of the moduli space $\mathcal{M}_g$ of genus g Riemann surfaces. As this boundary is generically not empty we arrive at the so called holomorphic anomaly, that is the failing of decoupling of ani-chiral fields.

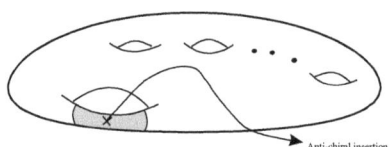

Figure 3.4: Contribution from first type degeneration

Taking the derivative of $\mathcal{F}^{(g)}$ with respect to $\bar{t}^i$ we obtain

$$\begin{aligned}\frac{\partial}{\partial \bar{t}^i}\mathcal{F}^{(g)} &= \int_{\mathcal{M}_g}[dm]\int d^2z\langle\oint_{C_z} G^+ \oint_{C'_z} \overline{G}^+ \bar{\phi}_{\bar{i}}(z)\prod_{a=1}^{3g-3}\int \mu_a G^- \int \bar{\mu}_{\bar{a}}\overline{G}^-\rangle_{\Sigma_g}\\
&= \int_{\mathcal{M}_g}[dm]\sum_{b,\bar{b}=1}^{3g-3}\langle\int \bar{\phi}_{\bar{i}}\int 2\mu_b T \int 2\bar{\mu}_{\bar{b}}\overline{T}\prod_{a\neq b}\int \mu_a G^-\prod_{\bar{a}\neq\bar{b}}\int \bar{\mu}_{\bar{a}}\overline{G}^-\rangle_{\Sigma_g}\\
&= \int_{\mathcal{M}_g}[dm]\sum_{b,\bar{b}=1}^{3g-3}4\frac{\partial^2}{\partial m_b \partial\bar{m}_{\bar{b}}}\langle\int \bar{\phi}_{\bar{i}}\prod_{a\neq b}\int \mu_a G^-\prod_{\bar{a}\neq\bar{b}}\int \bar{\mu}_{\bar{a}}\overline{G}^-\rangle_{\Sigma_g}. \quad (3.5.31)\end{aligned}$$

These lines demand some explanation. In the first line the contours C_z and C'_z are around the point z where the anti-chiral field $\bar{\phi}_{\bar{i}}$ is inserted. Moving these contours around in the Riemann surface, one picks up the commutators $\oint_{C_w} G^+ \cdot G^-(w) = 2T(w)$ and $\oint_{C_w} \overline{G}^+ \cdot \overline{G}^-(\bar{w}) = 2\overline{T}(\bar{w})$. The T and $\overline{T}$ insertions can then be converted to moduli derivatives $\frac{\partial}{\partial m}$ and $\frac{\partial}{\partial \bar{m}}$ as outlined in section 3.3.4. As a last step the Cauchy theorem can be used to convert the final line to an integral over the boundary of the moduli space $\mathcal{M}_g$. We will be rather brief in the following and refer the reader to [40] for a detailed explanation and calculation. There are basically two types of boundary components of the moduli space $\mathcal{M}_g$. The first type arises when a handle pinches off and the surface becomes a connected surface of genus $(g-1)$ with two punctures once the node is removed. The other type consists of surfaces of genus r and $(g-r)$ connected through a long thin tube. After removing the tube we obtain two disconnected surfaces of genus r and $(g-r)$ each with one puncture. The first type gives rise to the following anomaly term

$$\frac{1}{2}\overline{C}_{\bar{i}\bar{j}\bar{k}}e^{2K}G^{i\bar{j}}G^{k\bar{k}}D_j D_k \mathcal{F}^{(g-1)}, \quad (3.5.32)$$

where $\overline{C}$ denotes the antiholomorphic three-point function, K is the Kähler potential of the complex structure moduli space and $G_{i\bar{j}}$ is the Weil-Peterson metric. There is only a contribution to the integral for configurations where ϕ_i sits on the tube connecting the two nodes which is also the reason for the appearance of C_{ijk}. This configuration is depicted in figure 3.4.

The two covariant derivatives D_j and D_k are remnants of the removed node and correspond to the insertion of chiral fields at the two punctures. Note that the D_i are

3.6. MIRROR SYMMETRY

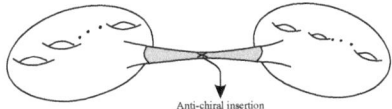

Figure 3.5: Contribution from second type degeneration

covariant derivatives with respect to the Weil-Peterson metric and the line bundle $\mathcal{L}$, i.e. one has
$$D_i = \partial_i - \Gamma_i - (2 - 2g)\partial_i K, \qquad (3.5.33)$$
where $\partial_i K$ is the connection on $\mathcal{L}$. The integration over the second type of component gives the following contribution to the anomaly
$$\frac{1}{2}\sum_{r=1}^{g-1} \overline{C}_{\bar{i}\bar{j}\bar{k}} e^{2K} G^{j\bar{j}} G^{k\bar{k}} D_j \mathcal{F}^{(r)} D_k \mathcal{F}^{(g-r)}. \qquad (3.5.34)$$

Again the two covariant derivatives arise from the one puncture on each of the disconnected surfaces where the anti-chiral field is inserted within the long tube as depicted in figure 3.5.

Together these two boundary integrals yield
$$\bar{\partial}_{\bar{i}} \mathcal{F}^{(g)} = \frac{1}{2}\overline{C}_{\bar{i}\bar{j}\bar{k}} e^{2K} G^{j\bar{j}} G^{k\bar{k}} \left(D_j D_k \mathcal{F}^{(g-1)} + \sum_{r=1}^{g-1} D_j \mathcal{F}^{(r)} D_k \mathcal{F}^{(g-r)} \right). \qquad (3.5.35)$$

This gives a recursion relation for $\mathcal{F}^{(g)}$ with respect to the genus g which we will solve up to an ambiguity in section 3.7.2. One should bear in mind that the above argumentation only goes through for Riemann surfaces with genus $g \geq 2$. In the case of $g = 1$ the holomorphic anomaly was computed in [46] and reads
$$\bar{\partial}_{\bar{k}} \partial_m \mathcal{F}^{(1)} = \frac{1}{2}\overline{C}_{\bar{k}}^{ij} C_{mij} - (\frac{\chi}{24} - 1)G_{\bar{k}m}, \qquad (3.5.36)$$
where we have defined $\overline{C}_{\bar{i}}^{kl} = e^{2K} G^{k\bar{k}} G^{l\bar{l}} \overline{C}_{\bar{i}\bar{k}\bar{l}}$.

3.6 Mirror Symmetry

As we have seen in our discussion of nonlinear Sigma-models on Calabi-Yau manifolds the two geometric deformations of the target space, namely complex structure and Kähler structure parameters, can be described from the conformal field theory point of view by so called marginal deformations. Kähler structure deformations can be parametrized by fields $\Phi_{(-1,1)}$ with $(U(1)_L, U(1)_R)$ charge $(-1,1)$ and complex structure deformations correspond to fields $\Phi_{(1,1)}$ with the charges $(U(1)_L, U(1)_R) = (1,1)$. These two kinds of conformal field theory operators just differ by a conventional sign of a $U(1)$ charge. However, from the geometric point of view we have the cohomology groups $H^1(M, \mathcal{T}_M)$ and $H^{2,1}(M)$ which are vastly different mathematical objects. This observation caused the authors of [30] and [27] to postulate the following correspondence: for each Calabi-Yau manifold M there is a second Calabi-Yau W such that the two are described by the *same conformal field theory* but with a *reversed* association of $H^{2,1}(M)$ and $H^1(M, \mathcal{T}_M)$ to conformal field theory marginal operators. This can be rephrased into the geometric conditions

$$h^{1,1}(M) = h^{2,1}(W), h^{2,1}(M) = h^{1,1}(W). \tag{3.6.1}$$

We see that this implies that the Hodge diamond for W is a *mirror reflection* through a diagonal axis of the Hodge diamond for M which is the reason why such a pair (M, W) is denoted by the term *mirror manifolds*.

3.6.1 Implications for the Topological String

As we have seen mirror symmetry identifies two manifolds M and W in a way such that the Kähler structure parameters of the one manifold get mapped to the complex structure variations of the other manifold. From the point of view of the topological string this observation appears to be very powerful. Indeed it can be shown that the partition function of the topological A model on M gets identifies with the one of the topological B model on W under the so called *mirror map*. Following this line of thought we see that calculations on the B model side, i.e. solving the holomorphic anomaly equations, already *contain* A model results and in particular the Gopakumar-Vafa invariants can be extracted from them by the use of the mirror map. In mathematical terms we thus have the following correspondence

$$\mathcal{F}_A^{(g)}(t_A, M) \equiv \mathcal{F}_B^{(g)}(t_B, W), \tag{3.6.2}$$

where the equivalence means the identification of the two free energies depending on A model parameters t_A and B model parameters t_B using the mirror map $t_A = t_A(t_B)$. We shall denote complex structure deformations by the collective coordinates $\underline{z}$ and Kähler deformation parameters by the collective coordinates $\underline{t}$, i.e. we have $t_A = t$ and $t_B = z$. We will make this correspondence more precise later on. As a first step we present the construction of the B model prepotential, i.e. $\mathcal{F}^{(0)}$, in terms of the periods of the holomorphic three-form. Being solutions to the Picard-Fuchs equations these periods can be calculated as a power series expansion around every point in moduli space but there also exist closed expressions for them.

For a general set of Picard-Fuchs equations, the solution space has dimension $h_3(W)$ and one obtains the following set of periods:

3.6. MIRROR SYMMETRY

$$\Pi(z) = \begin{pmatrix} \int_{B_1} \Omega \\ \vdots \\ \int_{B_{h^{2,1}+1}} \Omega \\ \int_{A^1} \Omega \\ \vdots \\ \int_{A^{h^{2,1}+1}} \Omega \end{pmatrix} = \begin{pmatrix} \mathcal{F}^{(0)} \\ \vdots \\ \mathcal{F}^{h_{2,1}} \\ X_0 \\ \vdots \\ X_{h^{2,1}} \end{pmatrix} = \begin{pmatrix} \omega_{2h^{2,1}+2} + \sum_{i=1}^{2h^{2,1}-1} c_i^0 \omega_i \\ \sum_{i=0}^{2h^{2,1}-1} c_i^1 \omega_i \\ \vdots \\ \omega_0 \\ \vdots \\ \omega_{h^{2,1}} \end{pmatrix}. \quad (3.6.3)$$

Here the ω_i are solutions to the Picard-Fuchs equations at the maximal unipotent monodromy point in moduli space and are organized as follows. ω_0 is the single solution starting with a constant, ω_k for $k = 1, \cdots, h^{2,1}$ denote single logarithmic solutions, ω_l for $l = h^{2,1} + 1, \cdots, 2h^{2,1} + 1$ denote solutions with two logarithms in the $\underline{z}$ and finally $\omega^{2h^{2,1}+2}$ is the single triple logarithmic solution. Using the fact that the ω_i are periods of the holomorphic three-form in a specific basis one sees that they can be transformed to the special geometry basis $(X^0, X^i, (\partial \mathcal{F}/\partial X^i), (\partial \mathcal{F}/\partial X^0))$ by a symplectic rotation.

On the other hand, working on the mirror side, the period vector $\Pi(t) = (1, t^i, \partial_i F, 2F - t^i \partial_i F)$ encodes Kähler deformations of the Calabi-Yau M with Kähler parameter t^i. Here F is the prepotential for the Kähler side and admits the formal large radius expansion

$$F = \frac{1}{6} \kappa_{ijk} t^i t^j t^k + \frac{1}{2} a_{ij} t^i t^j + b_i t^i + \frac{1}{2} c + F_{\text{inst}}. \quad (3.6.4)$$

For a specific choice of the constants c_i^j in (3.6.3) these two period vectors are related around $\text{Im}(t^i) \to \infty$ through $\Pi(z) = X^0 \Pi(t)$ with the choice

$$t^i(z) = \frac{\omega_i(z)}{\omega_0(z)}, \quad i = 1, \cdots, h^{2,1}. \quad (3.6.5)$$

From the periods we can calculate the triple couplings

$$C_{ijk} = \int_W \Omega \wedge \partial_i \partial_j \partial_k \Omega = D_i D_j D_k \mathcal{F}. \quad (3.6.6)$$

Note that the covariant derivatives w.r.t. the Weil-Petersen metric and the Kähler connection become ∂_{t_i} in the coordinates (3.6.5). This justifies the name flat coordinates for the t_i. Furthermore, we have the identification $K = -\log(X^0)$ with K being the Kähler potential associated to the prepotential.

As the higher genus prepotentials are sections of the line bundle $\mathcal{L}^{2-2g}$ we have the following identity relating the A to the B model free energy

$$F_A^g(\underline{t}) = (X^0)^{2g-2} \mathcal{F}^{(g)}(\underline{z}(\underline{t})). \quad (3.6.7)$$

The $F_A^g(\underline{t})$ are called Gromov-Witten potentials.

3.7 Solving the holomorphic anomaly equations

In this section we want to outline a method of solution of the B-model higher genus amplitudes using the holomorphic anomaly equations [40], the modular properties of the F^g [47, 49, 50] and boundary conditions in particular the gap conditions of [51].

3.7.1 The holomorphic limit

Before delving into the details of the direct integration method for solving the holomorphic anomaly equations let us first clarify the limit procedure we will use in order to perform actual computations. This limit procedure is called the *holomorphic limit* as it is an expansion around the base point $\bar{t} \to \infty$. In such an expansion all non-holomorphic quantities will become purely holomorphic. As the genus 0 free energy and correlation functions are purely holomorphic from the start, looking at the holomorphic anomaly equations (3.5.36), (3.5.35) we see that all anti-holomorphic dependence will come only from the metric $G_{i\bar{j}}$, the Christoffel symbol Γ_{ij}^k, and the Kähler potential K. Using the mirror map $\underline{t} = \underline{t}(\underline{z})$ and its inverse the holomorphic limits of these quantities are extracted to be

$$G_{\bar{i}j} \sim \frac{\partial t_k}{\partial z_l},$$
$$\Gamma_{ij}^k \sim \sum_m \frac{\partial z_k}{\partial t_m}(\underline{z})\partial_{z_i}\frac{\partial t_m(\underline{z})}{\partial z_j},$$
$$K \sim -\log(\omega_0), \tag{3.7.1}$$

where ω_0 is the solution starting with least possible power in the $\underline{z}$ [1]. All following calculations will make use of this limit, when calculating propagators and extracting the Gopakumar-Vafa invariants. However, it is important to keep in mind that all these quantities in fact exhibit anti-holomorphic dependence which is crucial for the procedure of direct integration to which we shall turn next.

3.7.2 Direct Integration

The method of direct integration relies on four key properties. The first is the fact that the $\mathcal{F}^{(g)}$ fulfill the holomorphic anomaly equations. The second is the fact that the $\mathcal{F}^{(g)}$ are modular invariant under the monodromy group Γ of the Calabi-Yau target space, which is a subgroup of $\text{Sp}(h_3, \mathbb{Z})$, and can be built from a finite polynomial ring of modular objects. In the large phase space these objects can be identified directly with modular forms under Γ [50], while the modular generators that appear below are obtained after a projection to the small phase space. The third important ingredient is the existence of a canonical antiholomorphic extension of the ring of modular forms to a ring of almost holomorphic forms, with the property that the appropriate covariant derivatives closes on

---
[1] At the large complex structure point this is the solution starting with $1 + \cdots$

3.7. SOLVING THE HOLOMORPHIC ANOMALY EQUATIONS

the almost holomorphic ring and that the antiholomorphic derivative in the holomorphic anomaly equation can be replaced by a derivative w.r.t the antiholomorphic generators of the almost holomorphic ring. The integration of the polynomials $\mathcal{F}^{(g)}$ w.r.t. the antiholomorphic generators leaves a holomorphic modular ambiguity, which is finitely generated over the smaller holomorphic ring. The final ingredients are physical boundary conditions at the discriminant components of the Calabi-Yau space, which determine the coefficients of the holomorphic modular ambiguity and allow only for a restricted class of modular objects in the rings, comparable to requiring restricted cusp behaviour for modular forms of $\Gamma_0 = Sl(2, \mathbb{Z})$.

Indeed the comparison to the classical theory of Γ_0 modular forms of elliptic curves [52] is very instructive. The ring of modular forms $\mathcal{M}_*[E_4, E_6]$ is here generated by the Eisenstein series E_4 and E_6. The covariant derivative is the Mass derivative acting on weight k modular forms by $D_k = \left(\frac{d}{2\pi i d\tau} - \frac{k}{4\pi \text{Im}(\tau)}\right)$. It does not close on $\mathcal{M}_*[E_4, E_6]$, but on the ring of almost holomorphic functions $\mathcal{M}^!\lbrack\hat{E}_2, E_4, E_6\rbrack$, where $\hat{E}_2$ is the anholomorphic extension of the second Eisenstein series $\hat{E}_2 = E_2 - \frac{3}{\pi \text{Im}(\tau)}$. The latter plays the role of the anholomorphic propagators in the formalism of [40]. Moreover a modular form w.r.t. Γ_0 of weight k fulfills a linear differential equation in the J-function of order $k+1$. This is the analog of the Picard-Fuchs equation and even if we know little about the modular objects of the Calabi-Yau group Γ it is possible to reconstruct them from the solutions of the Picard-Fuchs system. The totally invariant complex parameters $\underline{z}$ on the moduli space play here the rôle of the J-function. It should be noted that this is more than a formal analogy, because in certain local limits, as will be the subject of chapter 5, the formalism of the global Calabi-Yau space reduces to the one of a family of elliptic surfaces. For more details on modular forms we refer the reader to appendix B.

For the Calabi-Yau case the method of direct integration was developed in the work of Yamaguchi and Yau [47] for the one parameter models and extended in the work [49] to the multimoduli case.

Here we first follow the latter one as it is the more general construction and will give afterwards a short presentation of the method of [47] as this method is particularly suited for one-parameter models.

General case

The construction and the properties of the anholomorphic objects rely crucially on special geometry relation (3.1.41)

$$\bar{\partial}_{\bar{i}} \Gamma^k_{ij} = \delta^k_i G_{j\bar{i}} + \delta^k_j G_{i\bar{i}} - C_{ijl} \bar{C}^{kl}_{\bar{i}}, \qquad (3.7.2)$$

from which one can show [49, 47]

$$
\begin{aligned}
D_i S^{jk} &= \delta_i^j S^k + \delta_i^k S^j - C_{imn} S^{mj} S^{nk} + h_i^{jk}, \\
D_i S^j &= 2\delta_i^j S - C_{imn} S^m S^{nj} + h_i^{jk} K_k + h_i^j, \\
D_i S &= -\frac{1}{2} C_{imn} S^m S^n + \frac{1}{2} h_i^{mn} K_m K_n + h_i^j K_j + h_i, \\
D_i K_j &= -K_i K_j - C_{ijk} S^k + C_{ijk} S^{kl} K_l + h_{ij},
\end{aligned}
\quad (3.7.3)
$$

where

$$\partial_{\bar{i}} S^{ij} = \bar{C}_{\bar{i}}^{ij}, \quad \partial_{\bar{i}} S^j = G_{\bar{i}i} S^{ij}, \quad \partial_{\bar{i}} S = G_{\bar{i}i} S^i, \quad K_i = \partial_i K, \quad (3.7.4)$$

and h_i^{jk}, h_j^i, h_i and h_{ij} denote holomorphic functions. The propagators S^{ij}, S^i and S are obtained as solutions of the equations (3.7.4) up to holomorphic functions f_{kl}^i, f_{kl} and f:

$$
\begin{aligned}
S^{ij} &= (C_k^{-1})^{jl}((\delta_k^i \partial_l + \delta_l^i \partial_k) K + \Gamma_{kl}^i + f_{kl}^i), \\
S^i &= (C_k^{-1})^{il}(\partial_k K \partial_l K - \partial_k \partial_l K + f_{kl}^j \partial_j K) + f_{kl}), \\
S &= \frac{1}{2h^{11}} \left[(h^{1,1}+1) S^i - D_j S^{ij} - S^{ij} S^{kl} C_{jkl} \right] \partial_i (K + \log(|f|)/2) \\
&\quad + \frac{1}{2h^{1,1}} (D_i S^i + S^i S^{jk} C_{ijk}),
\end{aligned}
\quad (3.7.5)
$$

where the matrix C_k^{-1} is the inverse of the matrix $(C_k)_{ij} = C_{ijk}$. The relations (3.7.3) imply that the topological free energies $\mathcal{F}^{(g)}$ are polynomials in a finite set of non-holomorphic generators, namely the propagators S^{ij}, S^i, S and the Kähler derivatives K_i. To see this, note that equation (3.7.23) can be written in terms of these generators as

$$\partial_i \mathcal{F}^{(1)} = \frac{1}{2} C_{ijk} S^{jk} - (\frac{\chi}{24} - 1) K_i + A_i, \quad (3.7.6)$$

where the holomorphic ambiguity is encoded in the ansatz $A_i = \partial_i(\tilde{a}_j \log \Delta_j + \tilde{b}_j \log z_j)$. Rewriting the left hand side of equation (3.5.35) as

$$\bar{\partial}_{\bar{i}} \mathcal{F}^{(g)} = \bar{C}_{\bar{i}}^{jk} \frac{\partial \mathcal{F}^{(g)}}{\partial S^{jk}} + G_{\bar{i}i} \left(\frac{\mathcal{F}^{(g)}}{\partial K_i} + S^i \frac{\partial \mathcal{F}^{(g)}}{\partial S} + S^{ij} \frac{\partial \mathcal{F}^{(g)}}{\partial S^j} \right), \quad (3.7.7)$$

and assuming independence of the $\bar{C}_{\bar{i}}^{jk}$ and the $G_{\bar{i}i}$ gives

$$
\begin{aligned}
\frac{\partial \mathcal{F}^{(g)}}{\partial S^{ij}} &= \frac{1}{2} D_i D_j \mathcal{F}^{(g-1)} + \frac{1}{2} \sum_{r=1}^{g-1} D_i \mathcal{F}^{(g-r)} D_j \mathcal{F}^{(r)}, \\
0 &= \frac{\partial \mathcal{F}^{(g)}}{\partial K_i} + S^i \frac{\mathcal{F}^{(g)}}{\partial S} + S^{ij} \frac{\partial \mathcal{F}^{(g)}}{\partial S^j}.
\end{aligned}
\quad (3.7.8)
$$

Due to (3.7.3) and (3.7.6) the right hand side of these equations is always a polynomial in the generators (3.7.4). Therefore, it is straightforward to integrate the equations (3.7.8)

3.7. SOLVING THE HOLOMORPHIC ANOMALY EQUATIONS

which finally shows the polynomiality of the free energies. The last equation in (3.7.8) can be used to show that $\mathcal{F}^{(g)}$ becomes independent of the K_i in the redefined basis

$$\begin{aligned}
\tilde{S}^{ij} &= S^{ij}, \\
\tilde{S}^{i} &= S^{i} - S^{ij}K_j, \\
\tilde{S} &= S - S^{i}K_i + \frac{1}{2}S^{ij}K_iK_j, \\
\tilde{K}_i &= K_i,
\end{aligned} \qquad (3.7.9)$$

and one has $\partial \mathcal{F}^{(g)}/\partial \tilde{K}_i = 0$.

For practical calculations it is convenient to work in the basis of the tilted generators and we therefore rewrite the truncation relations (3.7.3) in terms of these as

$$\begin{aligned}
D_i\tilde{S}^{kl} &= \tilde{S}^l\delta_i^k + \tilde{S}^k\delta_i^l + \tilde{K}_j\tilde{S}^{jl}\delta_i^k + \tilde{K}_j\tilde{S}^{jk}\delta_i^l - C_{imn}\tilde{S}^{km}\tilde{S}^{ln} + h_i^{kl}, \\
D_i\tilde{S}^{k} &= 2\tilde{S}\delta_i^k + \tilde{K}_m\tilde{S}^m\delta_i^k - \delta_i^m\tilde{K}_m\tilde{S}^k - h_{im}\tilde{S}^{mk} + h_i^k, \\
D_i\tilde{S} &= -2\tilde{S}\tilde{K}_i - h_{im}\tilde{S}^m + \frac{1}{2}C_{imn}\tilde{S}^m\tilde{S}^n + h_i, \\
D_i\tilde{K}_j &= -\tilde{K}_i\tilde{K}_j - C_{ijk}\tilde{S}^k + h_{ij}.
\end{aligned} \qquad (3.7.10)$$

The holomorphic functions h_i^{kl}, h_i^k, h_i and h_{ij} are extracted from expansions of the above equations around the large complex structure point in moduli space and are valid after tensor transformation at every other point on the deformation space.

Method of Yamaguchi and Yau for one-parameter models

The basic idea is to introduce two sets of generators, given by

$$A_k = G^{z\bar{z}}\theta_z^k G_{z\bar{z}}, \quad B_k = e^{K(z,\bar{z})}\theta_z^k e^{-K(z,\bar{z})}, \qquad (3.7.11)$$

where $\theta_z = z\frac{d}{dz}$ and z is the only complex structure deformation parameter. A short calculation shows

$$\theta_z A_k = A_{k+1} - A_1 A_k, \quad \theta_z B_k = B_{k+1} - B_1 B_k. \qquad (3.7.12)$$

Noticing the relation $e^{-K(z,\bar{z})} = \langle \Omega(z), \bar{\Omega}(z) \rangle^2$, the fourth order Picard-Fuchs equation usually obtained for one-parameter models can be rewritten in terms of the B_k

$$B_4 = r_1(z)B_1 + r_2(z)B_2 + r_3(z)B_3 + r_4(z), \qquad (3.7.13)$$

where the $r_k(z)$ are rational functions.

Furthermore, there exists a similar relation for the A_k. As was shown in [47] A_2 is given by

---

[2]Here $\langle \Omega, \bar{\Omega} \rangle$ denotes the scalar product $-i\Pi^\dagger \Sigma \Pi$, where Π is the period vector and Σ the symplectic bilinear form

$$A_2 = -4B_2 - 2B_1(A_1 - B_1 - 1) + \theta_z \log(zC_{zzz})T_{zz} + r(z), \quad (3.7.14)$$

where T_{zz} is defined through the S^{zz} propagator

$$T_{zz} = -(zC_{zzz})S^{zz}, \quad (3.7.15)$$

and $r(z)$ is a holomorphic function to be specified later. Also the propagators are defined up to holomorphic functions f and v

$$\begin{aligned} S^{zz} &= \frac{1}{C_{zzz}}\left(2\partial \log(e^K|f|^2) - (G_{z\bar{z}}v)^{-1}\partial(vG_{z\bar{z}})\right) \\ &= -\frac{1}{zC_{zzz}}\left(2B_1 + 2\frac{\partial f}{f} + A_1 - \frac{\partial v}{v}\right). \end{aligned}$$

The choice for f and v is done such that the invariant combinations $e^K|f|^2$ and $G_{z\bar{z}}|v|^2$ remain finite around $z = 0$.

The rational function $r(z)$ is obtained by taking the holomorphic limit of both sides of equation (3.7.14) and making an appropriate Ansatz in terms of the discriminants.

The two equations (3.7.14) and (3.7.13) show that the θ_z-derivative acts within the ring generated by A_1, B_1, B_2 and B_3. More precisely, we have the property

$$\theta_z : \mathbb{C}(z)[A_1, B_1, B_2, B_3] \to \mathbb{C}(z)[A_1, B_1, B_2, B_3]. \quad (3.7.16)$$

Similarly, the action of the $\partial_{\bar{z}}$ derivative just adds two more generators to the above polynomial ring, namely $\partial_{\bar{z}}B_1$ and $\partial_{\bar{z}}A_1$. This is because, as was shown in [47], one has the following identities

$$\partial_{\bar{z}}B_2 = (1 + A_1 + 2B_1)\partial_{\bar{z}}B_1, \quad (3.7.17)$$

$$\partial_{\bar{z}}B_3 = (A_2 + 3B_1 + 3B_2 + 3A_1B_1 + 1)\partial_{\bar{z}}B_1. \quad (3.7.18)$$

The next step will be to show that rewriting the holomorphic anomaly equations allows us to rewrite the solutions in terms of polynomials in A_1, B_1, B_2 and B_3. In order to proceed we first introduce the quantities $P_n^{(g)}$ defined through

$$P_n^{(g)} = (z^3 C_{zzz})^{g-1} z^n D_z^n \mathcal{F}^{(g)} \quad (n = 0, 1, 2, \ldots). \quad (3.7.19)$$

Under the assumption that $\partial_{\bar{z}}A_1$, $\partial_{\bar{z}}B_1$ are independent the holomorphic anomaly equation

$$\partial_{\bar{z}}P^{(g)} = \frac{1}{2}\partial_{\bar{z}}(zC_{zzz}S^{zz})\left\{P_2^{(g-1)} + \sum_{r=1}^{(g-1)} P_1^{g-1}P_1^{(r)}\right\} \quad (3.7.20)$$

can be translated into

3.7. SOLVING THE HOLOMORPHIC ANOMALY EQUATIONS

$$0 = 2\frac{\partial P^{(g)}}{\partial A_1} - \left(\frac{\partial P^{(g)}}{\partial B_1} + \frac{\partial_{\bar{z}} B_2}{\partial_{\bar{z}} B_1}\frac{\partial P^{(g)}}{\partial B_2} + \frac{\partial_{\bar{z}} B_3}{\partial_{\bar{z}} B_1}\frac{\partial P^{(g)}}{\partial B_3}\right),$$

$$\frac{\partial P^{(g)}}{\partial A_1} = -\frac{1}{2}\left\{P_2^{g-1} + \sum_{r=1}^{g-1} P_1^{(g-r)} P_1^{(r)}\right\}.$$

This shows the polynomiality of the solutions. Performing the following variable change

$$u = B_1, \quad v_1 = 1 + A_1 + 2B_1, \quad v_2 = -B_1 - A_1 B_1 - 2B_1^2 + B_2,$$
$$v_3 = -B_1 - 2A_1 B_1 - 5B_1^2 - A_1 B_1^2 - 2B_1^3 + B_1 B_2 + B_3$$
$$- B_1(r(z) + T_{zz}\theta_z \log(zC_z zz)),$$

one can furthermore obtain $\frac{\partial}{\partial u}P^{(g)} = 0$ which reduces the number of independent variables to three. Notice that the above equations are generic for all kinds of one parameter models, once $r(z)$ is extracted from the truncation relation (3.7.14). The holomorphic anomaly equation can now be solved recursively with the initial data $P_3^{(0)} = 1$ and $P_1^{(1)}$, given by

$$P_1^{(1)} = \frac{1}{2}\left\{-A_1 - (2 + h^{1,1} - \frac{\chi}{12})B_1 - 1 - \frac{c_2 \cdot J}{12} - \frac{\theta_z(dis(z))}{6\,dis(z)}\right\}, \qquad (3.7.21)$$

where J is the Kähler form of the Calabi-Yau M.

However, the integration of the holomorphic anomaly still leaves us with the holomorphic ambiguity. The relation between the genus g free energy $\mathcal{F}^{(g)}$, the holomorphic ambiguity $f_g(z)$ and the polynomials $P^{(g)}$ is given by the following equation

$$\mathcal{F}^{(g)} = (z^3 C_{zzz})^{(1-g)} P^{(g)} + f_g(z). \qquad (3.7.22)$$

3.7.3 The holomorphic ambiguity

Having integrated the anomaly equations we remain with the task of fixing the integration constant which in fact is a meromorphic function defined on the whole of moduli space and called *holomorphic ambiguity*. This ambiguity is fixed by requiring certain boundary conditions at the boundary loci in moduli space. Taking the point of view of modular forms, the correct solution for the $\mathcal{F}^{(g)}$ has to admit singular expansions at cusps of the moduli space and be regular everywhere else. The degree and coefficient of the singularity will depend on the massless particle spectrum at the singular locus. In this thesis we will deal with the following types of singularities.

- Conifold locus: This divisor is universal and appears in the moduli space of all Calabi-Yau manifolds. As discussed in section 3.4.4, exactly one hypermultiplet becomes massless as the size of a S^2 shrinks to zero.

- Lense spaces: Lense spaces are cousins of conifolds where in the B model geometry the vanishing sphere is a three-sphere modded out by a discrete group. That is, at the divisor the space S^3/G, where $G \subset SU(2)$, shrinks to zero size. As is explained in [53] the number of hypermultiplets becoming massless is equal to the number of irreducible representations of G. We will come across examples of lense space singularities in chapter 4.

- ADE singularities: This type of singularity is common for K3 fibrations. It is characterized by singularities of type $\mathbb{C}^2/\mathbb{Z}_n$ fibred over a Riemann surface. The new feature of such a point in moduli space is the occurence of enhanced gauge symmetries as additional vector multiplets become massless. This phenomenon was discussed in section 2.2 and will be reviewed in more detail in chapter 6. The number of massless hypermultiplets is $2g$ for a genus g Riemann surface and that of massless vectormultiplets is 2 for a complex codimension 1 singularity.

- Landau-Ginzburg enhanced symmetry point: The Landau-Ginzburg enhanced symmetry point lies in a region of moduli space which is the analytic continuation of the geometric Kähler moduli space and thus the number of massless vector and hypermultiplets from vanishing cycles is zero. For hypersurfaces in weighted projective space there is a CFT description available at this point, see chapter 6.

Next, we will discuss the parametrization of the holomorphic ambiguity and boundary conditions at the singular loci.

Genus $g = 1$:

Equation (3.5.36) can be integrated straightforwardly and one obtains

$$\mathcal{F}^{(1)} = \frac{1}{2} \log \left[\exp \left[K(3 + h^{1,1} - \frac{\chi}{12}) \right] \det G_{i\bar{j}}^{-1} |f_1|^2 \right]. \quad (3.7.23)$$

f_1 is the holomorphic ambiguity arising form the integration and can be written in terms of the discriminant loci of the Calabi-Yau moduli space, i.e. $f = \prod_j \Delta_i^{a_i} \prod_{i=1}^{h^{2,1}} z_i^{b_i}$. All free parameters a_i, b_i are obtained through the limiting behavior of $\mathcal{F}^{(1)}$ near singularities. Canonical boundary conditions are given by the limit

$$\lim_{z_i \to 0} \mathcal{F}^{(1)} = -\frac{1}{24} \sum_{i=1}^{h^{2,1}} \log(z_i) \int_M c_2 J_i \quad (3.7.24)$$

as well as by the universal behavior at conifold singularities $a_{\text{con}} = -\frac{1}{12}$. The integral $\int_M c_2 J_i$ is performed on the mirror Calabi-Yau and M and J_i denote the Kähler forms associated to the different Kähler moduli.

3.7. SOLVING THE HOLOMORPHIC ANOMALY EQUATIONS

Genus $g \geq 2$:

As in the case of genus 1 there also arise holomorphic ambiguities at higher genus due to the anti-holomorphic derivative in (3.5.35). These ambiguities, denoted by f_g, are rational functions defined on the whole moduli space and transform as sections of $\mathcal{L}^{2-2g}$. One of the major challenges of topological string theory is to fix the ambiguity at each genus, after each integration step. This is done through using physical boundary conditions at the boundary divisors of the moduli space. In the case of compact Calabi-Yau manifolds one is dealing with several boundary divisors and many of them arise through a blow up of the moduli space and do not manifest themselves as singular loci of the Calabi-Yau hypersurface. A convenient way to see what is happening around these divisors and whether they have to be introduced in the holomorphic ambiguities for higher genera is to look at the behaviour of the genus 1 free energy $F_i^1(t_{i,N}, t_{i,T})$. Here, i is a label for the divisor in question and $t_{i,N}$, $t_{i,T}$ denote the flat coordinates normal as well as tangential to the divisor. In the work of Vafa [54] it is argued that the coefficient in front of the term logarithmic in $t_{i,N}$ counts the difference between hyper- and vector multiplets which become massless at the divisor Δ_i, i.e. we have the following expansion

$$F_i^1(t_{i,N}, t_{i,T}) = (n_H - n_V)\log(t_{i,N}) + \cdots . \qquad (3.7.25)$$

This allows us to constrain the form of the ambiguity for higher genera by demanding regularity at all divisors whose corresponding F^1-expansion does not come with a logarithmic term in the normal direction. This path of argumentation leads us to the following ansatz for the holomorphic ambiguities

$$f_g = \sum_{|I| \leq P_\infty(g)} a_I z^I + \sum_k \frac{\sum_{|I| \leq P_k(g) \cdot \deg \Delta_k} c_I^k z^I}{\Delta_k^{P_k(g)}}, \qquad (3.7.26)$$

where z^I is a short hand notation for $z_1^{i_1} z_2^{i_2} \cdots z_n^{i_n}$ and $|I| = i_1 + \cdots i_n$. Furthermore, $P_k(g)$ denotes the power of the boundary divisor Δ_k as a function of the genus g. Note that we also have terms which are polynomial in the z_i and therefore become singular around the locus Δ_∞ where $z_i \to \infty$.

The power of Δ_k in the denominator is fixed by the leading behaviour of F^g near the corresponding singularity. In the case of the conifold singularity the behaviour is of the form

$$F_c^g = \frac{c^{g-1} B_{2g}}{2g(2g-2)t_{c,N}^{2g-2}} + \mathcal{O}(t_c^0), \qquad (3.7.27)$$

where $t_{c,N} \to 0$ is a flat coordinate normal to the singularity locus. For more general singularities where n_H hypermultiplets and n_V vector multiplets become massless one expects the behaviour

$$F_s^g = (n_H - n_V)\frac{c^{g-1} B_{2g}}{2g(2g-2)t_{s,N}^{2g-2}} + \mathcal{O}(t_s^0), \qquad (3.7.28)$$

where $t_{s,N}$ is again the coordinate normal to the singularity locus. In order to extract the power of the discriminant component in the denominator of the ansatz one has to take into account the relation between Δ_s and $t_{s,N}$. In the case of the conifold discriminant this behaviour is a direct proportionality which is the reason why this discriminant appears to inverse powers of $2g-2$. In the case of the strong coupling discriminant we will be dealing in our examples the relation is $\Delta_s \sim t_{s,N}^2$ which leads to the ansatz

$$f_g = \ldots + \frac{\sum_{|I|\leq g-1} c_I^s z^I}{\Delta_s^{g-1}} + \ldots . \tag{3.7.29}$$

Formula (3.7.28) is a generalization of the Schwinger loop calculation performed in section 3.4.4 where one BPS hypermultiplet was running in the loop. As a BPS-vector multiplet contains one more fermion than a BPS-hypermultiplet there is a relative minus sign between the two loop calculations. Another way to see this is that in $N=4$ theories where one has no quantum corrections $N=2$ vector- and hypermultiplets are forming together one $N=4$ multiplet. Therefore, one expects that quantum corrections come with a sign difference.

However, the above argumentation leading to the result (3.7.28) goes only through once the theory is noninteracting and the calculation for the various BPS particles can be done separately. In an interacting theory several BPS states can form a bound state and then the calculation of the effective field theory becomes much more involved.

In the case of local Calabi-Yau manifolds with one conifold discriminant, reviewed in chapter 5, the ansatz (3.7.26) specializes to

$$f_g = \frac{A_g}{\Delta_{con}^{2g-2}}, \tag{3.7.30}$$

where the A_g are polynomials in z of degree $(2g-2)\cdot \Delta_{con}$. We find that the vanishing of subleading terms in (3.7.27) provides enough boundary conditions in order to fix the A_g and therefore the ambiguity completely. We claim that this is also the case in the compact examples, i.e. that the constants c_I^k are fixed completely by the leading behaviour near the corresponding singularity Δ_k. This claim is supported by calculations done in [3, 5], reviewed in chapters 4 and 6. However, once we are dealing with compact manifolds also terms of the form $a_I z^I$ appear in the ambiguity which become singular near the divisors $z_i = \infty$. The constant term in this series is always solved for by the constant map contribution to F^g at the point of large radius in moduli space

$$F^g = \frac{\chi B_{2g-2} B_{2g}}{4g(2g-2)(2g-2)!} + \mathcal{O}(e^{2\pi i t}). \tag{3.7.31}$$

The terms linear and of higher order in the z_i are connected to new physics becoming important at the divisors $z_i = \infty$ and/or their intersections. For hypersurfaces in weighted projective space one can prove that at the intersection point $z_1 = \cdots = z_{h_{2,1}} = \infty$ one has an exact CFT description and thus one can impose regularity on the amplitudes around this point.

Chapter 4

Grassmannian Calabi-Yau backgrounds

Mirror symmetry of Calabi-Yau manifolds has been understood to large extent for complete intersections or hypersurfaces in toric ambient space. However a huge and much less explored class of Calabi-Yau manifolds, with distinct low energy spectrum, can be realized in ambient spaces, which are defined by other homogeneous spaces like the Grassmannians $\mathbb{G}(k,n) = U(n)/(U(k) \times U(n-k))$. We shall denote denote Calabi-Yau manifolds, which are complete intersections in Grassmannians as "Grassmannian Calabi-Yau manifolds" and such which are realized as complete intersections in toric spaces as "toric Calabi-Yau manifolds". As discussed in section 3.2.2 from the point of view of the 2-d linear σ-model description of the ambient space the difference is that the former have $U(1)^r$ gauge symmetries, while the latter have non-abelian $\prod_k U(N_k)$ gauge symmetries.

This chapter reviews the results of reference [3] which analyzes the topological string on Grassmannian Calabi-Yau manifolds. In our presentation we will concentrate on those models which contain new distinct physics and will refer to [3] for other models with similar behaviour.

4.1 Calabi-Yau complete intersections in Grassmannians

In this section we introduce the Calabi-Yau intersections in Grassmannian, calculate their topological data and review the mirror construction of [59].

4.1.1 Topological invariants of the manifolds

Compact Calabi-Yau manifolds M can be constructed by considering complete intersections in Kähler ambient spaces with positive Chern class. The first Chern class of the complete intersections is controlled by the adjunction formula and we can chose appropriate degrees of the complete intersection constraints so that $c_1(\mathcal{T}_M) = 0$. We will calculate the topological data of M by basic algebraic geometry. All necessary tools are reviewed in [21, 60].

We restrict to complete intersections in smooth Grassmannians. In this way one finds 5 complete intersections M with $h^{1,1} = 1$. The ambient space will be denoted as $\mathbb{G}(k,n) = (U(k) \times U(n-k))$, where $U(n)$ are the unitary groups. For the complete intersection we use the notation

$$(\mathbb{G}(k,n)\|d_1,\ldots,d_l)_\chi^{h^{1,1}}. \tag{4.1.1}$$

Here the degrees d_i of the Calabi-Yau intersection are given w.r.t. the principal canonical bundle Q of the Grassmannian, see below. In addition we give the Euler number χ as subscript and the Picard number $h^{1,1}$ as superscript. Of course, $h_{3,0} = 1$, $h_{k,0} = 0$ for $k = 1, 2$ and $h^{2,1} = -\frac{\chi}{2} + h^{1,1}$. Together with Poincareé and Hodge duality this fixes all Hodge numbers of M. All necessary topological data, which fix the topological type of M, are calculated below using the Schubert calculus.

Let us first give a closed expression for the Chern classes of Grassmannians following Borel and Hirzebruch in [60]. Their method is based on an identification of Chern classes with elementary symmetric polynomials or combinations of them, which we will summarize here.

Let $S\{x_1,\cdots,x_l\}$ denote the set of elementary symmetric polynomials in the variables $x_1,\cdots,x_l$. Then the integral homology $H_*(\mathbb{G}(k,n),\mathbb{Z})$ of the Grassmannian can be identified with the quotient

$$S\{x_1,\cdots,x_{n-k}\} \otimes S\{x_{n-k+1},\cdots,x_n\}/I, \tag{4.1.2}$$

where I is the ideal generated by the symmetric power series in $x_1,\cdots,x_n$ without constant term. Now, in this representation, the closed formula for the total Chern class reads

$$c(\mathbb{G}(k,n)) = \prod_{i=1}^{n-k}(1-x_i)^n \prod_{1 \leq i \leq j \leq n-k}(1-(x_i-x_j)^2)^{-1}. \tag{4.1.3}$$

Practically, in order to calculate the Chern classes, substitute each x_l by hx_l and make a series expansion in h. Then, the i's Chern class is given by the coefficient of h^i which can be expressed in terms of elementary symmetric polynomials σ_r, $r \leq i$ in $x_1,\cdots,x_{n-k}$. For example, we have

$$\begin{aligned}c_1(\mathbb{G}(k,n)) &= -n\sigma_1, \\ c_2(\mathbb{G}(k,n)) &= \left(\binom{n}{2}+n-k-1\right)\sigma_1^2 + k\sigma_2.\end{aligned} \tag{4.1.4}$$

The formula for the first Chern class shows that $-\sigma_1$ is a positive generator of $H^2(\mathbb{G}(k,n),\mathbb{Z})$. Next, note that σ_r is (up to a possible sign) the r-th Chern class of the canonical principal

4.1. CALABI-YAU COMPLETE INTERSECTIONS IN GRASSMANNIANS

$U(n-k)$-bundle Q over $\mathbb{G}(k,n)$ and as such represents the class of a hyperplane section. We have $\sigma_1 = -c_1(Q)$, $\sigma_2 = c_2(Q)$, $\sigma_3 = -c_3(Q)$, ...

Finally, we are ready to write down the total Chern class of Calabi-Yau complete intersections $(\mathbb{G}(k,n)\|d_1,\ldots,d_l)_\chi^{h^{1,1}}$, $l = k(n-k) - 3$, $d_1 + \cdots + d_l = n$:

$$c((\mathbb{G}(k,n)\|d_1,\ldots,d_l)_\chi^{h^{1,1}}) = \frac{c(\mathbb{G}(k,n))}{(1+d_1 c_1(Q))\cdots(1+d_l c_1(Q))}. \quad (4.1.5)$$

Denoting by H the hyperplane σ_1, the topological invariants $\chi(M)$, $c_2(M)\cdot H$, H^3 can be expressed through intersection numbers of the Grassmannian $\mathbb{G}(k,n)$. As an example, we review the calculation of the Euler number. The Gauss-Bonnet formula gives $\int_M c_3(M) = \chi$. Now, using the adjunction formula, this integral can be expressed through an integral over the whole Grassmannian

$$\chi(M) = \int_M c_3(M) = \int_{\mathbb{G}(k,n)} c_3(M) \prod_{i=1}^{l} d_i H = \int_{\mathbb{G}(k,n)} c_3(M) \prod_{i=1}^{l} d_i c_1(Q). \quad (4.1.6)$$

Similarly, the other topological invariants are given by

$$c_2(M)\cdot H = \int_{\mathbb{G}(k,n)} c_2(M) c_1(Q) \prod_{i=1}^{l} d_i c_1(Q), \quad (4.1.7)$$

$$H^3 = \int_{\mathbb{G}(k,n)} c_1(M)^3 \prod_{i=1}^{l} d_i c_1(Q). \quad (4.1.8)$$

As all Chern classes of M are expressed through Chern classes of Q, which are Poincare dual to the Schubert cycles of the Grassmannian, all invariants can at the end be expressed through intersection numbers of Schubert cycles. These numbers can then be calculated utilizing the Schubert calculus and Pieri's formula. Denoting by σ_a the special Schubert cycle given by the indices $a = (a, 0, \cdots, 0)$ and by $\sigma_{\underline{b}}$ a general Schubert cycle with indices $\underline{b} = (b_1, \cdots, b_k)$, Pieri's formula reads

$$\sigma_a \cdot \sigma_{\underline{b}} = \sum_{\substack{b_i \leq c_i \leq b_{i-1} \\ \sum c_i = a + \sum b_i}} \sigma_{\underline{c}}. \quad (4.1.9)$$

Note that in the above formula the index c_1 must always be greater or equal to b_1. For further details we refer to [21].

We have performed the above steps and list the result for our Calabi-Yau complete intersections in the Appendix.

4.1.2 Plücker embedding

In order to describe the mirror of the complete intersections in Grassmannians it is useful to have an embedding of the Grassmannian into the projective space. The Plücker map provides such an embedding. It simply sends a k-plane $\Lambda = \mathbb{C}\{v_1, \cdots, v_k\} \subset \mathbb{C}^n$ to the multivector $v_1 \wedge \cdots \wedge v_k$.

Explicitly, in terms of the basis $\{e_I = e_{i_1} \wedge \cdots \wedge e_{i_k}\}_{\#I=k}$ for $\wedge^k \mathbb{C}^n$, this map is given by the data

$$p : \mathbb{G}(k,n) \to \mathbb{P}(\wedge^k \mathbb{C}^n) = \mathbb{P}^{\binom{n}{k}-1}, \qquad (4.1.10)$$

$$\Lambda \mapsto [\cdots, |\Lambda_I|, \cdots], \qquad (4.1.11)$$

where the $|\Lambda_I|$ are the determinants of all the $k \times k$ minors of Λ_I of a matrix representative of Λ.

To describe this embedding algebraically we need to find a set of equations which cut out the Grassmannian in $\mathbb{P}^{\binom{n}{k}-1}$, i.e. which define conditions on a multivector $\Lambda \in \wedge^k V$ to be of the form

$$\Lambda = v_1 \wedge \cdots \wedge v_k. \qquad (4.1.12)$$

Some calculations show that this is equivalent to demanding

$$(i(\Xi)\Lambda) \wedge \Lambda = 0, \qquad (4.1.13)$$

for all $\Xi \in \wedge^{k-1} V$. Here, the map $i(\Xi)\Lambda$ is defined by

$$\langle i(\Xi)\Lambda, v \rangle = \langle \Xi, \Lambda \wedge v \rangle \qquad (4.1.14)$$

for all $v \in V$.

Now, a Calabi-Yau complete intersection is obtained by choosing hypersurfaces of appropriate total degree in $\mathbb{P}^{\binom{n}{k}-1}$, such that their intersection with $\mathbb{G}(n,k)$ is a nonsingular Calabi-Yau space.

4.1.3 Mirror Construction

A mirror construction for the above type of Calabi-Yau spaces was given in [59]. Here, we will only sketch the method introduced there which is based on conifold transitions.

Let M be a Calabi-Yau complete intersection described by the Grassmannian $\mathbb{G}(k,n)$ and hyperplanes H_i. As was shown by Sturmfels [61] a flat deformation of $\mathbb{G}(k,n)$ in its Pluecker embedding leads to a Gorenstein toric Fano variety $P(k,n) \subset \mathbb{P}^{\binom{n}{k}-1}$. Now, denote by M_0 the intersection of $P(k,n)$ with generic hypersurfaces H_i. This manifold has a locus of conifold singularities which come from the singularities of $P(k,n)$. Resolving these by restriction of a small toric resolution of singularities in $P(k,n)$ one obtains a second Calabi-Yau M^*. M^* is a complete intersection in a toric manifold and as such its

4.2. PICARD-FUCHS EQUATIONS FOR ONE-PARAMETER MODELS

mirror construction is known. The remaining task is to find an appropriate specialization of the toric mirror W^* for M^* to a conifold W_0 whose small resolution provides the mirror W of M. This task was performed in [59] for the manifolds we will be dealing with in this paper.

The above steps can be summarized in the following graph:

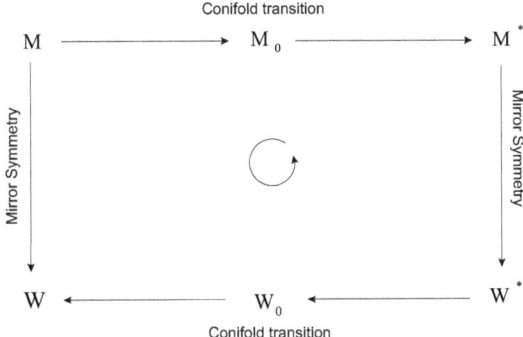

Figure 4.1: Mirror construction for Grassmannian Calabi-Yau manifolds.

4.2 Picard-Fuchs equations for one-parameter models

As the third homology group $H_3(W)$ of a mirror Calabi-Yau W with one complex structure parameter is four-dimensional the Picard-Fuchs equations governing the Hodge structure of such models are 4th order differential operators.

In order to obtain period expressions one therefore has to solve an ordinary fourth order differential equation. Solutions at different points in moduli space are constructed by first transforming the Picard-Fuchs operator to the relevant point and then solve its *indicial equation*. Applying the Picard-Fuchs operator to expressions of the form

$$z^\rho (1 + \sum_{n=1}^\infty a_n z^n), \qquad (4.2.1)$$

the indicial equation arises as the vanishing condition for the lowest order term in z. The roots $(\rho_1, \rho_2, \rho_3, \rho_4)$ will then determine the leading power behaviour of the four period solutions at the relevant point in moduli space. In the following we will stress some properties of the roots ρ_i and refer to [62] for a more complete treatment.

Picard-Fuchs operators with degenerate roots of the indicial equation admit regular singular points and signal the occurrence of logarithms. At the point of maximal unipotent monodromy in moduli space the roots are given by

$$(\rho_1, \rho_2, \rho_3, \rho_4) = (0, 0, 0, 0). \tag{4.2.2}$$

From this it follows that only one of the solutions is polynomial in z, all others being logarithmic. Another important point in moduli space is the *conifold point*. It is characterized by the indicials $(\rho_1, \rho_2, \rho_3, \rho_4) = (0, 1, 1, 2)$. This suggests that the third solution will be the product of the second one with a logarithm plus a polynomial part, i.e.[1]

$$\omega_2^c = \omega_1^c \log(z - z_c) + g(z), \tag{4.2.3}$$

where z_c is a conifold point. Going around z_c once ω_2^c is replaced by $\omega_2^c(z) + 2\pi i \omega_1^c$. This has the following geometric interpretation where we shall follow the notations and conventions of [65]. Locally at the conifold the defining function of the singularity is given by

$$f : \mathbb{C}^4 \to \mathbb{C}, \quad f(x, y, z, t) = x^2 + y^2 + z^2 + t^2. \tag{4.2.4}$$

The fibre F_s of f over $s \in \mathbb{C} - \{0\}$ is called the *Milnor fibre* and can be identified with the cotangent bundle to the sphere $\{(x, y, z, t) \in \mathbb{R}^4 | x^2 + y^2 + z^2 + t^2 = s\}$. Therefore one sees that a three-sphere S^3 is shrinking to zero size as $s \to 0$. Let δ be the homology class of this sphere and ϵ the covanishing cycle in the dual group $H_3^{cl}(F_s, \mathbb{Z})^2$ One has

$$H_3(F_s, \mathbb{Z}) = \mathbb{Z}\delta, \quad H_3^{cl}(F_s, \mathbb{Z}) = \mathbb{Z}\epsilon, \quad \langle \delta, \epsilon \rangle = 1, \tag{4.2.5}$$

where $\langle \cdot, \cdot \rangle$ denotes the intersection number of two homology cycles. With these definitions the monodromy operator of transporting a homology cycle in the integral homology lattice Λ once around the singularity z_c is determined to be

$$S_{1,\delta} : \Lambda \to \Lambda, \quad S_{1,\delta}(\gamma) = \gamma - \langle \delta, \gamma \rangle. \tag{4.2.6}$$

This is known as the classical *Picard-Lefschetz* formula. However, we can generalize it to the following case. Consider a discrete subgroup $G \subset SU(2) = S^3$ which acts linearly on $\mathbb{R}^4$ and by complexification on $\mathbb{C}^4$, leaving the defining function f of the singularity invariant. This way one can define a function $g : \mathbb{C}^4/G \to \mathbb{C}$ such that the Milnor fibre $G_s = g^{-1}(s)$ contains a vanishing cycle S^3/G. Denote the homology class of this vanishing cycle by $d \in H_3(G_s, \mathbb{Z})$. Then it follows by comparing the volume of d and δ that the Picard-Lefschetz formula gets modified to

$$\gamma \mapsto \gamma - |G|\langle d, \gamma \rangle d. \tag{4.2.7}$$

A comment is at order here. The generalized Picard-Lefschetz operator $S_{\lambda,\beta}(\alpha) = \alpha - \lambda\langle\beta,\alpha\rangle\beta$ is not of finite order, i.e. we have $S_{\lambda,\beta} \circ S_{\lambda',\beta} = S_{\lambda+\lambda',\beta}$. This finishes our

---
[1] We shall refer to periods at the conifold with an index c
[2] cl denotes homology with closed support.

discussion of the conifold and its cousins for one-parameter models. A further singular point in moduli space is usually the point at infinity $z_\infty := \frac{1}{z} = 0$. Here the solutions can admit fractional roots which signals a local $\mathbb{Z}_n$ orbifold for common denominators n.

Last but not least let us describe the connection between the solutions of the Picard-Fuchs equation and the periods of the A model mirror in the special geometry basis. To recover the period integrals over the basis $\{A^k, B_k\}$ from the solutions of the Picard-Fuchs equations we use special geometry and the typical degeneration of the periods at the point of maximal unipotent monodromy. First we note that the X^k serve as homogenous coordinates for the space of complex structures. Recall that $\mathcal{F}^{(0)}(X^k) := \frac{1}{2} X^k \mathcal{F}_k(X^k)$ is homogenous of degree 2 in X^k and $\mathcal{F}_k = \partial_{X^k} \mathcal{F}^{(0)}$. At the point of maximal unipotent monodromy we have

$$\vec{\Pi} = \begin{pmatrix} \int_{B_1} \Omega \\ \int_{B_2} \Omega \\ \int_{A^1} \Omega \\ \int_{A^2} \Omega \end{pmatrix} = \begin{pmatrix} \mathcal{F}_0 \\ \mathcal{F}_1 \\ X_0 \\ X_1 \end{pmatrix} = \omega_0 \begin{pmatrix} 2F^0 - t\partial_t F^0 \\ \partial_t F^0 \\ 1 \\ t \end{pmatrix} = \begin{pmatrix} \omega_3 + c\,\omega_1 + e\,\omega_0 \\ -\omega_2 - a\,\omega_1 + c\,\omega_0 \\ \omega_0 \\ \omega_1 \end{pmatrix},$$
(4.2.8)

where ω_0 is the unique power series solution and ω_k are solutions, which behave like $\omega_0(z)\log(z)^k$ at infinity. The Frobenius method gives a canonical basis of these solutions. $t = \frac{\omega_1}{\omega_0}$ is the mirror map and in terms of the latter the prepotential looks as follows

$$F^0 = -\frac{\kappa}{3!} t^3 - \frac{a}{2} t^2 + ct + \frac{e}{2} + f_{inst}(q) \,,$$
(4.2.9)

where $\kappa = H^3$, $c = \frac{1}{24} \int_M c_2 \wedge H$, $e = \frac{\zeta(3)\chi(M)}{(2\pi i)^3}$ and $a = \frac{1}{2} \int_M i_* c_1(H) \wedge H$. All these numbers are calculated on M using the formalism in section 4.1.1 and they fix the integral symplectic basis on W completely.

4.3 The Grassmannian Calabi-Yau $(\mathbb{G}(2,5)\|1,1,3)^1_{-150}$

This Calabi-Yau manifold is obtained as a complete intersection of hypersurfaces in the Grassmannian $\mathbb{G}(2,5)$ as described in section (4.1). In our special case the Plücker embedding is an embedding of $\mathbb{G}(2,5)$ into $\mathbb{P}^9$ and equations (4.1.14) take the form

$$\begin{aligned} z_{23}z_{45} - z_{24}z_{35} + z_{25}z_{34} &= 0, \\ z_{13}z_{45} - z_{14}z_{35} + z_{15}z_{34} &= 0, \\ z_{12}z_{45} - z_{14}z_{25} + z_{15}z_{34} &= 0, \\ z_{12}z_{35} - z_{13}z_{25} + z_{15}z_{23} &= 0, \\ z_{12}z_{34} - z_{13}z_{24} + z_{14}z_{23} &= 0. \end{aligned}$$
(4.3.1)

Now, the Calabi-Yau $(\mathbb{G}(2,5)\|1,1,3)^1_{-150}$ is defined to be a smooth 3-dimensional Calabi-Yau complete intersection of 3 hypersurfaces of degrees $1, 1$ and 3 in $\mathbb{P}^9$ with $\mathbb{G}(2,5)$. A calculation shows that we have $h^{1,1} = 1$, $h^{2,1} = 76$ and $\chi(M) = -150$.

4.3.1 Picard-Fuchs differential equation and the structure of the moduli space

The Picard-Fuchs operator for this model was extracted in [59] and is given by:

$$\begin{aligned}\mathcal{P} &= -18z - 360z^2 + (-147z - 2106z^2)\theta + (-444z - 3969z^2)\theta^2 \\ &\quad + (-594z - 2916z^2)\theta^3 + (1 - 297z - 729z^2)\theta^4,\end{aligned} \quad (4.3.2)$$

where $\theta = z\frac{d}{dz}$. As one can read off, the discriminant is given by dis$(z) = 1 - 297z - 729z^2$. The Yukawa coupling can be extracted from the Picard-Fuchs operator and its normalization is determined by the intersection number H^3 as explained in appendix A.

$$C_{zzz} = \frac{15}{z^3(1 - 11 \cdot 3^3 z - 3^9 z^2)}. \quad (4.3.3)$$

We expect the solutions to develop logarithmic singularities around the points dis$(\alpha_i) = 0$, $i \in \{1, 2\}$. These indeed occur as can be seen from the possible solutions ρ_i of the indicial equation:

$$(\rho_1, \rho_2, \rho_3, \rho_4) = (0, 1, 1, 2). \quad (4.3.4)$$

The monodromy behaviour at these points together with the expression for F^1 (see (4.3.10)) suggests that they are conifold-points of the moduli space. As discussed in section (3.4.4) at these points non-perturbative RR-states become massless and integrating them out leads to singularities in the effective action calculated by the topological string. We will use the gap condition discussed in section (3.4.3) to put restrictive bounds on the holomorphic ambiguity.

Another special point in our particular moduli space is the point at infinity. Here the Picard-Fuchs-operator develops the following indices: $(\rho_1, \rho_2, \rho_3, \rho_4) = (\frac{1}{3}, \frac{2}{3}, \frac{4}{3}, \frac{5}{3})$. The $\mathbb{Z}_3$-symmetry at this point suggests that it is the enhanced symmetry point of a particular Landau-Ginzburg orbifold model. Putting regularity conditions on topological string free energies at this point gives us another bound on the holomorphic ambiguity and the resulting Gopakumar-Vafa invariants will give us a consistency check whether our regularity assumption was justified.

Finally, the structure of the singularities can be summarized in the following table

z	0	α_1	α_2	∞
ρ_1	0	0	0	1/3
ρ_2	0	1	1	2/3
ρ_3	0	1	1	4/3
ρ_4	0	2	2	5/3

(4.3.5)

4.3.2 $g = 0$ and $g = 1$ Gopakumar-Vafa invariants

In this section we summarize the calculations of the genus zero and one Gopakumar-Vafa invariants for the Grassmannian. We will solve the Picard-Fuchs equation around the point $z = 0$ corresponding to maximal unipotent monodromy and obtain the mirror map at this point.

The normalized regular solution and the linear-logarithmic solution are

$$\left.\begin{array}{rl} \omega_0(z) &= 1 + 18z + 1710z^2 + 246960z^3 + 43347150z^4 + \cdots \\ \omega_1(z) &= \log \omega_0(z) + 75z + \frac{16497}{2}z^2 + 1257046z^3 + \frac{907324065}{4}z^4 + \cdots \end{array}\right\} \quad (4.3.6)$$

The complexified Kähler modulus is defined through $2\pi i t = \frac{\omega_1(z)}{\omega_0(z)}$ and the q-expansion of the z-coordinate takes the following form:

$$z = q - 75q^2 + 1539q^3 - 60073q^4 + \cdots, \quad (4.3.7)$$

where $q := e^{2\pi i t}$.

Now, we are able to determine the quantum corrected Yukawa coupling $K_{ttt}(t)$ at $z = 0$. It is given by

$$\left(\frac{1}{\omega_0(z)}\right)^2 C_{zzz} \left(\frac{dz}{dt}\right)^3 = 15 + 540q + 100980q^2 + 16776045q^3 + 2873237940q^4 + \cdots. \quad (4.3.8)$$

From these Yukawa couplings we can obtain the Gromov-Witten potential

$$K_{ttt}(t) = \left(q\frac{d}{dq}\right)^3 F_0(t). \quad (4.3.9)$$

The genus one invariants are obtained by taking the holomorphic limit of (3.7.23)

$$\mathcal{F}^{(1)}(z) = \frac{1}{2}\log\left\{\left(\frac{1}{\omega_0(z)}\right)^{3+h^{1,1}-\frac{\chi}{12}}\left(\frac{dz}{dt}\right) dis(z)^{-\frac{1}{6}} z^{c-1-\frac{c_2 \cdot H}{12}}\right\}, \quad (4.3.10)$$

where we determine $c = 0$ through the boundary behavior (3.7.24). As both zeros of the discriminant describe conifold points, it appears with factor $-1/12$ in the logarithm.

Using the mirror map $z = z(q)$ we finally obtain the genus one Gromov-Witten potentials

$$F^1(t) = \mathcal{F}^{(1)}(z(q)). \quad (4.3.11)$$

4.3.3 Higher genus free energies

In this section we compute higher genus free energies by exploiting the method of Yamaguchi and Yau outlined in section 3.7.2. The holomorphic limits at certain points in the moduli space are calculated. Imposing boundary conditions on the holomorphic ambiguity we can fix the holomorphic ambiguity at least up to genus 5.

Holomorphic ambiguity and boundary conditions

Requiring regularity for $F^g(t)$ at $z = 0$ and $z = \infty$, we parameterize the holomorphic ambiguity through the Ansatz

$$f_g(z) = a_0 + a_1 z + \cdots + a_{2g-2} z^{2g-2} + \frac{c_0 + c_1 z + \cdots + c_{4g-5} z^{4g-5}}{dis(z)^{2g-2}}. \tag{4.3.12}$$

From this we see that the total number of unknown parameters is $6(g-1) + 1$ and grows linearly in g.

Boundary conditions may be given through the effective 4d action as discussed in section 3.7.3, but also, in some cases, geometrical considerations can be of use. For example, we can utilize the first few n_d^g in the Gopakumar-Vafa expansion of the Gromov-Witten potential once they are known through geometrical calculations. Usually, one puts the lower degree Gopakumar-Vafa invariants n_d^g to zero as they count the number of genus g holomorphic curves in the Calabi-Yau. Once one knows that the n_d^g are vanishing up a certain degree for a specific genus g, then one knows that they must be zero at least up to the same degree for genus $g + 1$. This knowledge one can impose as boundary condition for the Gromov-Witten potentials. As boundary conditions from physical considerations are far more restrictive for higher genus calculations we will concentrate on these in this paper. In order to fix the ambiguity we evaluate the Gromov-Witten potentials at special points on the moduli space, where the physics is sufficiently well understood.

Expansion around the conifold points

In order to make use of the gap condition we have to compute the holomorphic limit around each conifold singularity. We denote the conifold singularity by c, i.e. in our case c stands for either $\alpha_1 = 1/54(-11 - 5\sqrt{5})$ or $\alpha_2 = 1/54(-11 + 5\sqrt{5})$. In the following we will obtain a normalized set of solutions of the Picard-Fuchs differential equation. From the index structure around the conifold (4.3.5), the existence of a logarithmic solution can be deduced. Furthermore, we have solutions which start with s^i ($s = (z-c)$, $i = 0, 1, 2$) which we will denote by $\omega_i^c(s)$. We normalize the logarithmic solution $\log(s)\omega_1^c(s) + \mathcal{O}(s^1)$ by requiring $\omega_1^c(s) = s + \mathcal{O}(s^2)$. The solution corresponding to the index $\rho_4 = 2$ is normalized to be of the form $\omega_2^c(s) = s^2 + \mathcal{O}(s^3)$. A suitable linear combination with $\omega_1^c(s)$ and $\omega_2^c(s)$ allows us to choose the solution for the index $\rho_1 = 0$ to be of the form

$$\omega_0^c(s) = 1 + \mathcal{O}(s^3). \tag{4.3.13}$$

The mirror map can be now specified to be

$$k_t t_c = \frac{\omega_1^c(s)}{\omega_0^c(s)}, \tag{4.3.14}$$

where k_t is a constant which for the moment we can set to one.

We solve the Picard-Fuchs equations over the ring $\mathbb{Q}[\alpha]/dis(\alpha)$ and obtain the following results for the periods and the mirror maps

4.3. THE GRASSMANNIAN CALABI-YAU $(\mathbb{G}(2,5)\|1,1,3)^1_{-150}$

$$\omega_0^\alpha(s) = 1 + \frac{81}{250}(435709 + 1060776\alpha)s^3 + \mathcal{O}(s^4)$$
$$\omega_1^\alpha(s) = s - \frac{3}{50}(3709 + 9126\alpha)s^2 + \frac{3}{25}(446957 + 1088046\alpha)s^3 + \mathcal{O}(s^4)$$
(4.3.15)
$$s(t_\alpha) = t_\alpha - \frac{3}{50}(3709 + 9126\alpha)t_\alpha^2 + \frac{3}{50}(770597 + 1875852\alpha)t_\alpha^3 + \mathcal{O}(t_\alpha^4) \quad (4.3.16)$$

In order to regain the solutions around the points α_i, $i \in \{1, 2\}$ one has to substitute α by α_i. For more details about this method see [63].

The holomorphic limits around the conifold points are obtained by making the replacements

$$A_1(s+c, \bar{s}+\bar{c}) \to (s+c)\frac{d}{ds}\log\frac{dt_c}{ds}, \quad B_k \to \frac{1}{\omega_0^c(s)}((s+c)\frac{d}{ds})^k \omega_0^c(s) \qquad (4.3.17)$$

in the expansions of the free energies in terms of the A and B generators.

Specializing the gap condition to one-parameter models we now obtain

$$F_c^g(t_c) = (\omega_0(s))^{2g-2}\mathcal{F}_c^{(g)}(s) = \frac{\text{const.}}{t_c^{2g-2}} + \mathcal{O}(t_c^0), \qquad (4.3.18)$$

for $g \geq 2$. This provides us with $(2g-2) - 1$ equations which are vanishing conditions for the coefficients of $\frac{1}{t_c^i} (1 \leq i \leq 2g-3)$. Actually, the condition is even stronger as there exists a choice of the constant k_t under which in all higher genus expansions the leading term is of the form $\frac{|B_{2g}|}{2g(2g-2)}\frac{1}{t_c^{2g-2}}$.

It is interesting to have a look at this gap structure in the expansions of Gromov-Witten potentials once the holomorphic ambiguity is fixed completely,

$$F_\alpha^2(t_\alpha) = \frac{41 - 12276\alpha}{874800 t_\alpha^2} + \frac{-14874743 + 3442099023\alpha}{36450000} + O(t_\alpha),$$
$$F_\alpha^3(t_\alpha) = -\frac{5(-15005 + 4493016\alpha)}{4821232752 t_\alpha^4} + \mathcal{O}(t_\alpha^0). \qquad (4.3.19)$$

Again, substitute α by α_i to obtain the solutions around the specific vanishing point of the discriminant.

Expansion around the orbifold point

The index structure (4.3.5) of the Picard-Fuchs operator suggests that the point at infinity is a $\mathbb{Z}_3$ orbifold point. Therefore, we have to impose regularity of the free energies at this point in the moduli space. To obtain the topological limits we follow a path of argumentation presented in [63]. Let x be the coordinate at infinity, i.e. $x = \frac{1}{z}$. Then

98 CHAPTER 4. GRASSMANNIAN CALABI-YAU BACKGROUNDS

we can define $\tilde{\mathcal{F}}^{(g)}(x,\bar{x})$ to be the solutions of the BCOV equation in x-coordinates with initial conditions $\tilde{\mathcal{F}}_1^{(1)}(x,\bar{x})$ and $\tilde{\mathcal{F}}_3^{(0)} = D_x D_x D_x \tilde{\mathcal{F}}^{(0)}(x,\bar{x})$. On the other hand these initial conditions are related by

$$\tilde{\mathcal{F}}_3^{(0)}(x,\bar{x}) = C_{xxx}(x) = C_{zzz}\left(\frac{1}{x}\right)\left(\frac{dz}{dx}\right)^3 = \mathcal{F}_3^{(0)}\left(\frac{1}{x},\frac{1}{\bar{x}}\right)\left(\frac{dz}{dx}\right)^3. \tag{4.3.20}$$

From this we can infer that $\tilde{\mathcal{F}}^{(g)}(x,\bar{x})$ and $\mathcal{F}^{(g)}(z,\bar{z})$ are in the same coordinate patch of a trivialization of the line bundle $\mathcal{L}$, which again gives

$$\tilde{\mathcal{F}}^{(g)}(x,\bar{x}) = \mathcal{F}^{(g)}\left(\frac{1}{x},\frac{1}{\bar{x}}\right). \tag{4.3.21}$$

Therefore, the topological limit at infinity is simply obtained by setting $\tilde{\mathcal{F}}^{(g)}(x,\bar{x}) = \mathcal{F}^{(g)}(A_1(\frac{1}{x},\frac{1}{\bar{x}}), B_k(\frac{1}{x},\frac{1}{\bar{x}}), \frac{1}{x})$ and taking the limits

$$A_1\left(\frac{1}{x},\frac{1}{\bar{x}}\right) = \left(\frac{dz}{dx}\frac{d\bar{z}}{d\bar{x}}G^{x\bar{x}}\right)(-\theta_x)\left(\frac{dx}{dz}\frac{d\bar{x}}{d\bar{z}}G_{x\bar{x}}\right) \to -\left(\frac{dx}{dt_\infty}\right)\theta_x\left(\frac{dt_\infty}{dx}\right) - 2 \tag{4.3.22}$$

$$B_k\left(\frac{1}{x},\frac{1}{\bar{x}}\right) = e^{\tilde{K}(x,\bar{x})}(-\theta_x)^k e^{-\tilde{K}(x,\bar{x})} \to \frac{1}{\omega_0^\infty(x)}(-\theta_x)^k \omega_0^\infty(x), \quad (k=1,2,3) \tag{4.3.23}$$

where $\omega_0^\infty(x), \omega_1^\infty(x)$ and $t_\infty(x) = \frac{\omega_1^\infty(x)}{\omega_0^\infty(x)}$ are the periods and mirror map at infinity.

So in order to proceed we have to calculate these quantities first. From the index structure we have the following set of solutions, $\omega_0^\infty(x) = x^{1/3} + \mathcal{O}(x^{4/3})$, $\omega_1^\infty(x) = x^{2/3} + \mathcal{O}(x^{5/3})$, $\omega_2^\infty(x) = x^{4/3} + \mathcal{O}(x^{7/3})$ and $\omega_3^\infty(x) = x^{5/3} + \mathcal{O}(x^{8/3})$. Using a linear combination with $\omega_2^\infty(x)$ we can fix the first solution to be of the form

$$\omega_0^\infty(x) = x^{1/3} + \mathcal{O}(x^{7/3}). \tag{4.3.24}$$

Furthermore, the second solution can be fixed by taking a linear combination with the third solution to

$$\omega_1^\infty(x) = x^{2/3} + \mathcal{O}(x^{8/3}). \tag{4.3.25}$$

With these choices the relevant solutions are given by

$$\begin{aligned}
\omega_0^\infty(x) &= x^{1/3} + \frac{x^{7/3}}{131220} - \frac{67}{51018336}x^{10/3} + \mathcal{O}(x^{13/3}), \\
\omega_1^\infty(x) &= x^{2/3} - \frac{2}{45927}x^83 - \frac{467}{55801305}x^{11/3} + \mathcal{O}(s^{14/3}), \\
x &= t_\infty^3 - \frac{11}{102060}t_\infty^9 + \frac{12599}{595213920}t_\infty^{12} + \mathcal{O}(t_\infty^{15}).
\end{aligned} \tag{4.3.26}$$

Using these data and the holomorphic limit discussed above we obtain the following Gromov-Witten potentials

$$F_\infty^2(t_\infty) = \frac{\frac{41031}{160} + a_2}{t_\infty^4} + \frac{\frac{1367}{80} + a_1}{t_\infty} + \mathcal{O}(t_\infty),$$

$$F_\infty^3(t_\infty) = \frac{\frac{22453281}{1600} + a_4}{t_\infty^8} + \frac{\frac{4572543}{3200} + a_3}{t_\infty^5} + \frac{-\frac{121464319}{567000} + a_2 + \frac{73 a_4}{229635}}{t_\infty^2}$$

$$= +\mathcal{O}(t_\infty). \qquad (4.3.27)$$

As the orbifold point is a conformal field theory point and thus has to be regular, we see that demanding the vanishing of the coefficients of inverse powers of t_∞ gives us g conditions on the parameters of the holomorphic ambiguity.

Counting the number of boundary conditions from the orbifold and conifold points one notices that they are not yet enough to fix the ambiguity completely. This is no problem for lower genera as the vanishing of lower degree Gopakumar-Vafa invariants gives us enough conditions to fix all free parameters. On the other hand, as mentioned earlier, our example shows that there are not enough boundary conditions to solve the model up to genus infinity.

4.4 Other Models

We have analysed three other Calabi-Yau complete intersections in Grassmannians, namely $(\mathbb{G}(2,5)\|1,2,2)^1_{-120}$, $(\mathbb{G}(3,6)\|1^6)^1_{-96}$ and $(\mathbb{G}(2,6)\|1,1,1,2)^1_{-116}$. All three admit interesting new features and share common properties with the model analysed previously. In particular, we have found a lense space point in the moduli space of the second model.

4.4.1 $(\mathbb{G}(2,5)\|1,2,2)^1_{-120}$

The topological data of this Calabi-Yau are given by $\chi = -120$, $h^{2,1} = 61$, $h^{1,1} = 1$, $c_2 \cdot J = 68$. The Picard-Fuchs operator which was obtained in [59] admits the following index structure

z	0	α_1	α_2	∞
ρ_1	0	0	0	1/2
ρ_2	0	1	1	1/2
ρ_3	0	1	1	3/2
ρ_4	0	2	2	3/2

(4.4.1)

and the Yukawa coupling is determined to be

$$C_{zzz} = \frac{20}{z^3(1 - 11 \cdot 2^4 z - 2^8 z^2)}. \qquad (4.4.2)$$

100   CHAPTER 4. GRASSMANNIAN CALABI-YAU BACKGROUNDS

The indicial structure at the points α_1 and α_2 suggests that these points are conifold points and indeed the expansion $F^1(t_c) = \frac{1}{12}\log(t_c) + \cdots$ gives the universal conifold coefficient $\frac{1}{12}$. However, the from the index structure at infinity we see that logarithmic solutions are to appear. So although the fractional indices suggest a $\mathbb{Z}_2$ orbifold point, this point will be a hybrid of orbifold type and conifold type singularities.

For the solutions around the conifold points we choose exactly the same normalization as in the case of $(\mathbb{G}(2,5)\|1,1,3)^1_{-150}$.

Looking at the point at infinity, we see that there are two logarithmic solutions. In order to obtain the mirror map only the first two solutions ω_0^∞ and ω_1^∞ are needed. They are of the form

$$\begin{aligned}\omega_0^\infty &= x^{1/2} + \mathcal{O}(x^{5/2}),\\ \omega_1^\infty &= \log(x)x^{1/2} + \mathcal{O}(x^{9/2}),\end{aligned} \qquad (4.4.3)$$

and we take the mirror map to be of the form $t = \frac{\omega_1^\infty(x)}{\omega_0^\infty(x)}$.

With these conventions we calculate the expansions of the free energies around the singular points of the moduli space. We find the same gap conditions as in the case of $(\mathbb{G}(2,5)\|1,1,3)^1_{-150}$ around the two conifolds. The point at infinity turns out to be a regular point as we have to impose regularity on the Gromov-Witten potentials in order to obtain integral Gopakumar-Vafa numbers. We list the genus 2 and 3 expansions around this point

$$\begin{aligned}F^2_\infty(t_\infty) &= \frac{5^{1/4}(136+3a_2)}{48\sqrt{3}t_\infty^{1/4}} + (a_1 + \frac{-119464 - 4047a_2}{32000}) + \mathcal{O}(t_\infty),\\ F^3_\infty(t_\infty) &= \frac{\sqrt{5}(\frac{1024}{3}+a_4)}{768\sqrt{t_\infty}} + \frac{-28849664 + 144000a_3 - 36423a_4}{460800\sqrt{3}5^{3/4}t_\infty^{1/4}} + \mathcal{O}(t_\infty).\end{aligned} \qquad (4.4.4)$$

As one can see regularity restrictions give us $g-1$ boundary conditions on the ambiguity.

4.4.2 $(\mathbb{G}(3,6)\|1^6)^1_{-96}$

This Calabi-Yau has the topological data $\chi = -96$, $h^{2,1} = 49$, $h^{1,1} = 1$, $c_2 \cdot J = 84$. The Picard-Fuchs operator given in [59] admits the following index structure

z	0	α_1	α_2	∞
ρ_1	0	0	0	4/3
ρ_2	0	1	1	1
ρ_3	0	1	1	1
ρ_4	0	2	2	5/4

(4.4.5)

4.4. OTHER MODELS

The Yukawa coupling is given by

$$C_{zzz} = \frac{28}{z^3(1 - 26 \cdot 2^2 z - 27 \cdot 2^4 z^2)}. \tag{4.4.6}$$

The point at infinity admits one logarithmic solution which corresponds to a vanishing cycle and it appears that it also admits some orbifold features. The mirror map is given by $t = \frac{\omega_1^\infty(x)}{\omega_0^\infty(x)}$, where

$$\begin{aligned} \omega_0^\infty &= x^{3/4} + \mathcal{O}(x^{7/4}), \\ \omega_1^\infty &= x + \mathcal{O}(x^2). \end{aligned} \tag{4.4.7}$$

An interesting feature of this model is the fact that the two vanishing points of the discriminant, although having the same Picard-Fuchs-indices, behave differently when we analyze the Gromov-Witten potentials. In particular, the genus 1 Gromov-Witten potential of this model is

$$F^1(z) = \frac{1}{2} \log \left\{ \left(\frac{1}{\omega_0(z)} \right)^{3+h^{1,1} - \frac{\chi}{12}} \left(\frac{dz}{dt} \right) (-1+z)^{-\frac{1}{3}} (-1+64z)^{-\frac{1}{6}} z^{-1 - \frac{c_2 \cdot H}{12}} \right\}. \tag{4.4.8}$$

This suggests that the point $z = 1$ is not an ordinary conifold point but rather a lense space point, that is a point, where a cycle $\mathcal{C}$(for example S^3) modded by a group G shrinks to zero size. In the case of $\mathcal{C} = S^3$ G is a discrete subgroup of $SU(2)$ and the resulting space S^3/G has fundamental group G. Spaces of this form where investigated in [64], where the number of BPS states admitted by such cycles was calculated. There it was argued that the number of D-brane bound states which are BPS is equal to the number of irreducible representations of G and their mass is given by the formula $M_i = \mu d_i/G$ where μ is the size of the unmodded cycle and d_i is the dimension of the ith irreducible representation of G. Comparing this with the genus one free energy of the topological string one finds

$$F^1 = \sum_i -\frac{1}{12} \log(M_i) = \sum_i -\frac{1}{12} \log(\mu d_i/G). \tag{4.4.9}$$

In our particular example this is

$$F^{(1)} = -\frac{1}{12} \log(t_{1/64}) - \frac{2}{12} \log(t_1). \tag{4.4.10}$$

Using the identification $t_1 = \mu/2$ we find from the above formula that the group G must be $\mathbb{Z}_2$. This also shows that two hypermultiplets are becoming massless at $z = 1$.

Our result is supported by the monodromy calculations made in [65]. There it was found that the monodromy matrix at the point $z = 1$ is of Picard-Lefschetz form $S_{\lambda,v}$, where $\lambda = 2$ which shows that this point is not an ordinary conifold point.

Higher genus calculations show that the ordinary gap condition holds at $z = 1/64$ which is to be expected as this point is a conifold point. On the other hand the gap condition has to be slightly modified around $z = 1$. If we assume that the two hypermultiplets becoming massless are not interacting the modification to the leading term of the higher genus Gromov-Witten potential reads as follows

$$F_1^g(t_1) = 2\frac{|B_{2g}|}{2g(2g-2)}\frac{1}{\mu^{2g-2}} + \mathcal{O}(t_1^0) = 2\frac{|B_{2g}|}{2g(2g-2)}\frac{1}{2^{2g-2}}\frac{1}{t_1^{2g-2}} + \mathcal{O}(t_1^0). \qquad (4.4.11)$$

This is exactly what we observe.

It remains to be discussed the point at infinity. It admits a gap-like structure as can be seen for example from the genus 4 expansion

$$F_\infty^4(t_\infty) = \frac{7}{240\, t_\infty^6} + \frac{101797151}{11010048000}t_\infty^2 + \mathcal{O}(t_\infty^3). \qquad (4.4.12)$$

4.4.3 $(\mathbb{G}(2,6)\|1,1,1,1,2)^1_{-116}$

This manifold is characterized by the data $\chi = -116$, $h^{2,1} = 59$, $h^{1,1} = 1$, $c_2 \cdot J = 76$. The structure of the solutions of the Picard-Fuchs operator is the following

z	0	α_1	α_2	∞
ρ_1	0	0	0	1/2
ρ_2	0	1	1	2/3
ρ_3	0	1	1	4/3
ρ_4	0	2	2	3/2

(4.4.13)

The Yukawa coupling is given by

$$C_{zzz} = \frac{42}{z^3(1 - 65z - 64z^2)}. \qquad (4.4.14)$$

The conifold locus is treated as usual. The mirror map at $z = \infty$ is obtained by taking the ratio of the first two periods. They are of the form

$$\begin{aligned}\omega_0^\infty &= x^{1/2} + \mathcal{O}(x^{5/2}),\\ \omega_1^\infty &= x^{2/3} + \mathcal{O}(x^{5/3}).\end{aligned} \qquad (4.4.15)$$

Now, our calculations show that the gap condition holds at the conifold locus. Furthermore, the point at infinity at first sight seems to be a regular orbifold point with $\mathbb{Z}_6$-symmetry and indeed this seems to be the case up to genus 3. But at genus 4 we find that the expansion of the Gromov-Witten potential around this point is singular. In particular we find

$$F^4_\infty(t_\infty) = \frac{-\frac{8606402923}{164640} + a_6}{t_\infty^{18}} + \frac{-\frac{500305024099}{49787136} + a_5 - \frac{10}{63}a_6}{t_\infty^{12}}$$
$$+ \frac{-\frac{443407050538901893}{179412923289600} + a_4 - \frac{20}{189}a_5 + \frac{831575}{54486432}a_6}{t_\infty^6} + \mathcal{O}(t_\infty^0), \quad (4.4.16)$$

before fixing the ambiguity and

$$F^4_\infty(t_\infty) = \frac{2}{2187\, t_\infty^6} + \frac{108172361}{131681894400} + \mathcal{O}(t_\infty), \quad (4.4.17)$$

after having fixed the ambiguity.

4.4.4 $(\mathbb{G}(2,7)\|1^7)^1_{-98}$

This manifold is characterized by the data $\chi = -98$, $h^{2,1} = 50$, $h^{1,1} = 1$, $c_2 \cdot J = 84$. The structure of the solutions of the Picard-Fuchs operator is the following

z	0	α_1	α_2	α_3	3	∞
ρ_1	0	0	0	0	0	1
ρ_2	0	1	1	1	1	1
ρ_3	0	1	1	1	3	1
ρ_4	0	2	2	2	4	1

(4.4.18)

We see that the Picard-Fuchs differential operator has the property of maximal degeneration at both $z = 0$ and $z = \infty$. It was found in [67] that the expansion about $z = 0$ corresponds to the Kähler moduli of the Grassmannian Calabi-Yau $M = (\mathbb{G}(2,7)\|1^7)^1_{-98}$, and the expansion about $z = \infty$ to that of a Pfaffian Calabi-Yau M'. In [63] the instanton calculations for this model were extended up to genus 5 and we confirm their results for low genus.

4.5 5d black hole entropy

As outlined in section 2.3 solving the topological string to all genus is important to study black holes in five and four dimensions [66]. E.g. for five dimensional black holes in $N=2$ supergravity with spin m and charge $Q \in H_2(M,\mathbb{Z})$ there is a microscopic prediction (see equation (2.3.11) of section 2.3)

$$S(Q,M) = \log\left(\sum_{r=0}^{\infty} \binom{2r+2}{m+r+1} n^g(Q)\right), \quad (4.5.1)$$

where $n^g(Q)$ are the BPS degeneracies that we can calculate now on the Grassmannian Calabi-Yau manifolds. Note that we have $h_2(M, \mathbb{Z}) = 1$ and we denote the charge Q from now on, like in the previous parts of this chapter, by the degree d of the curve w.r.t. the divisor H.

The macroscopic calculations are valid in the limit $d \gg m$. The $N = 2$ supergravity action has higher derivative corrections of the form $F_+^{2g-2} R_+^2$, where F_+ and R_+ are the selfdual part of the graviphoton field strength and the curvature respectively. The evaluation of Wald's entropy formula for the uncorrected action yields

$$S_0 = 2\pi \sqrt{Q^3 - m^2}, \qquad (4.5.2)$$

where $\mathcal{Q}$ is the graviphoton charge which is determined by the attractor mechanism in terms of the triple intersection $\mathcal{Q} = \left(\frac{2}{9H^3}\right)^{\frac{1}{3}} d$. The first correction yields

$$S_1 = \frac{\pi c_2 \cdot H}{8} \left(\frac{6}{H^3}\right)^{\frac{1}{3}} \sqrt{d^3 - m^2} \left(\frac{1}{d} + \frac{m^2}{3d^4}\right). \qquad (4.5.3)$$

Higher $F_+^{2g-2} R_+^2$ corrections are expected to be of the form $S_g = \int_M c_3 \mathcal{Q}^{\frac{3}{2}-g}$. We can now make a large d expansion of the total macroscopic entropy $S = b_0 d^{\frac{3}{2}} + b_1 d^{\frac{1}{2}} + \mathcal{O}\left(\frac{1}{d^{\frac{1}{2}}}\right)$ and compare the coefficients b_i with the corresponding expansion of (4.5.1). So far this can be done only numerically as the exact asymptotic of the BPS states is not known. It is notable that the range of the topological data, which determine the b_i take more extreme values for the Grassmannians than for the toric varieties. In particular $c_2 \cdot H$ and the triple intersection H^3 take the highest values for Grassmannian Calabi-Yau. This is very useful for comparing the semiclassical and the microscopic description of black holes along the lines of [66]. Indeed we find that the microscopic entropy the Richardson transforms converge within 4 % to the expected value of the macroscopic calculation. For reference we show one plot for the extreme value of $H^3 = 42$ in the appendix.

4.6 Summary of the models

Let us summarize at this point the results we have obtained for the various models analyzed. We find that the model $(\mathbb{G}(3,6) \| 1^6)^1_{-96}$ has a conifold at $z = \frac{1}{64}$ and a lense space $S^3/\mathbb{Z}_2$ shrinking at $z = 1$. We find that at the lense space singularity the analysis of the leading terms is exactly as predicted in [53] and that in addition there is a full gap structure in the subleading terms. The physical interpretation is that the two BPS states do not interact and in particular do not form light bound states. This model has also at t_∞ a branch point of order 12, a single logarithmic solution and a full gap structure.

The models $(\mathbb{G}(2,5) \| 1, 1, 3)^1_{-150}$, $(\mathbb{G}(2,5) \| 1, 2, 2)^1_{-120}$ are regular at $t_\infty = 0$ at least to genus 5. The first has regular solutions, which hints at a CFT with a $\mathbb{Z}_3$ automorphism at $t_\infty = 0$. In this model the BPS invariant $n_6^4 = 5$ has been checked geometrically by Sheldon Katz, who found also the vanishing of the BPS invariants for the other model in accord with Castelnouvos Theory.

4.6. SUMMARY OF THE MODELS

The model $(\mathbb{G}(2,5)\|1,2,2)^1_{-120}$ has two logarithmic solutions and a branch point of order 2. It is conceivable that higher F^g are not regular at $t_\infty = 0$.

The model $(\mathbb{G}(2,6)\|1,1,1,1,2)^1_{-116}$ has two different conifolds with a full gap structure. At the point $t_\infty = 0$ it has regular solutions with a $\mathbb{Z}_6$ branching. Curiously we find that the integrality of the BPS require that it has singular behavior in the F^g for $g > 3$.

For the Rodland example $(\mathbb{G}(2,7)\|1^7)^1_{-98}$, which has two points of maximal unipotent monodromy we confirm the analysis of [63] for low genus.

Chapter 5
Local Calabi-Yau backgrounds

String theory on non-compact Calabi-Yau geometries is relevant for the construction of 4d supersymmetric theories decoupled from gravity as outlined in section 2.2.2. Moreover it provides simple examples for important concepts of string theory in nontrivial geometrical backgrounds, as e.g. the behavior of the amplitudes under topology change of the background geometry. From the point of view of the 2d σ model these geometries correspond to σ models without superpotential and with both, positive and negative $U(1)$ charges. Exploring the topological sector has been especially fruitful in providing examples of large N-dualities connecting topological string theory on these backgrounds to 3d Chern-Simons theory and matrix models.

This chapter reviews the results of reference [4] where it was shown that the holomorphic anomaly equations become integrable for local Calabi-Yau geometries whose mirror contains a Riemann surface of at least genus 1. We restrict ourselves to the models $\mathbb{F}_0$ and $\mathbb{F}_1$.

5.1 Local Mirror Symmetry

The term local mirror symmetry refers to mirror symmetry for non-compact Calabi-Yau manifolds. Examples for the A-model geometry are the canonical line bundle $\mathbb{K}_S = \mathcal{O}(-K_S) \to S$ over a Fano surface [1] S. The compact part of the B-model geometry is in this case given by a family of elliptic curves and a meromorphic differential. Using toric geometry as below an infinite set of examples of non-compact three-folds can be constructed. They have a partial overlap with the $\mathbb{K}_S$ cases namely $S = \mathbb{P}^1 \times \mathbb{P}^1$ or $S = \mathbb{P}^2$ and blow-ups thereof $S = \mathbb{BP}_1^2, \mathbb{BP}_2^2, \mathbb{BP}_3^2$. The mirror geometry are Riemann surfaces with a meromorphic differential, whose genus is given by the number of closed meshes in the degeneration locus in the base of symplectic fibration, where two S^1's degenerate. For early applications of local mirror symmetry to BPS state counting and

---
[1] Simpler examples involve line bundles over a complex curve such as $\mathcal{O}(2(g-2)+k) \oplus \mathcal{O}(-k) \to \mathcal{C}_g$ [68] or manifolds M, which are given by a toric tree diagrams of the degeneration locus that correspond to genus 0 mirror curves.

geometric engineering of gauge theories see [69] and [13] respectively. For a systematic formulation see [70][71][72]. Below we give a very short review of the techniques.

5.1.1 The local A-model

The A-model geometry of a non-compact toric variety is given by a quotient

$$M = (\mathbb{C}^{k+3} - Z)/G, \qquad (5.1.1)$$

where $G = (\mathbb{C}^*)^k$ [74]. On the homogeneous coordinates $x_i \in \mathbb{C}$ the group G acts like $x_i \to \mu_\alpha^{Q_i^\alpha} x_i$, $\alpha = 1,\ldots,k$ with $\mu_\alpha \in \mathbb{C}^*$, $Q_i^\alpha \in \mathbb{Z}$. Here Z is the Stanley-Reissner ideal, which has to be chosen so that the above quotient M exists as a variety[2]. The standard example is $\mathbb{P}^n = (\mathbb{C}^{n+1} - \{0\})/(\mathbb{C}^*)$, with $Q_i^1 = 1$, $i = 1,\ldots,n$. We denote generically by $\mathcal{S}$ the compact part of M.

As explained in 3.2.2 M can also be viewed as the vacuum field configuration of a 2d gauged linear $(2,2)$ supersymmetric σ model. The coordinates $x_i \in \mathbb{C}$, $i = 1,\ldots,k+3$ are the vacuum expectation values of chiral superfields transforming as $x_i \to e^{iQ_i^\alpha \epsilon_\alpha} x_i$, $Q_i^\alpha \in \mathbb{Z}$, $\epsilon_\alpha \in \mathbb{R}$, $\alpha = 1,\ldots,k$ under the gauge group $U(1)^k$. The vacuum field configuration are the equivalence classes under the gauge group, which fulfill in addition the D-term constraints

$$D^\alpha = \sum_{i=1}^{k+3} Q_i^\alpha |x_i|^2 = r^\alpha, \quad \alpha = 1,\ldots,k\ . \qquad (5.1.2)$$

The r^α are the Kähler parameters $r^\alpha = \int_{C_\alpha} \omega$, where ω is the Kähler form and C_α are curves spanning the Mori cone, which is dual to the Kähler cone. $r^\alpha \in \mathbb{R}_+$ defines the Kähler cone. For M to be well defined, field configurations for which the dimensionality of the gauge orbits drop have to be excluded. This corresponds to the choice of Z. In string theory r^α is complexified to $T^\alpha = r^\alpha + i\theta^\alpha$ with $\theta^\alpha = \int_{C_\alpha} B$, where B is the NS B-field, while in the gauged linear σ-model the θ^α are the θ-angles of the $U(1)^k$ gauge group.

One can always describe M by a completely triangulated fan. In this case the Q_i^α are linear relations between the points spanning the fan. A basis of such relations, which corresponds to a Mori cone can be constructed from a complete triangulation of the fan. Z likewise follows combinatorially from the triangulation, see the examples[3].

The Calabi-Yau condition $c_1(\mathcal{T}_M) = 0$ holds if and only if[4]

$$\sum_{i=1}^{k+3} Q_i^\alpha = 0, \qquad \alpha = 1,\ldots,k. \qquad (5.1.3)$$

---

[2] We assume that M is simplicial, or that a simplicial subdivision in coordinate patches exists.

[3] Often there are many possible triangulations, which correspond to different phases of the model see section 3.2.2 and [75], e.g. Kähler cones connected by flopping a $\mathbb{P}^1$. The union of the cones define by all triangulations is called the secondary fan.

[4] Physically these are the conditions that the chiral $U(1)_A$ anomaly cancels in the gauged linear σ-model see 3.2.2.

5.1. LOCAL MIRROR SYMMETRY

Note from (5.1.2) that negative Q_i lead to non-compact directions in M, so that by (5.1.3) all toric Calabi-Yau manifolds M are necessarily non-compact. To summarize, toric non-compact A-model geometries will be defined by suitably chosen charge vectors $Q_i^\alpha \in \mathbb{Z}$.

5.1.2 The local B-model

In the following we will describe the non-compact mirror W following [71, 13]. Let $w^+, w^- \in \mathbb{C}$ and $x_i =: e^{y_i} \in \mathbb{C}^*$, $i = 1, \ldots, k+3$ are homogeneous coordinates[5], i.e. equivalence classes subject to the $\mathbb{C}^*$ action

$$x_i \mapsto \lambda x_i, \quad i = 1, \ldots, k+3, \quad \lambda \in \mathbb{C}^* .\tag{5.1.4}$$

The mirror W is defined from the charge vectors Q_i^α by the exponentiated D-term constraints

$$(-1)^{Q_0^\alpha} \prod_{i=1}^{k+3} x_i^{Q_i^\alpha} = z_\alpha, \quad \alpha = 1, \ldots, k .\tag{5.1.5}$$

and the general equation

$$w^+ w^- = H = \sum_{i=1}^{k+3} x_i .\tag{5.1.6}$$

The Calabi-Yau condition (5.1.3) ensures the compatibility of (5.1.5) with (5.1.4). Using the latter two equations to eliminate variables x_i in (5.1.6) H can be parameterized by two variables $x = \exp(u), y = \exp(v) \in \mathbb{C}^*$ and the defining equations of the mirror geometry W becomes

$$w^+ w^- = H(x, y; z_\alpha),\tag{5.1.7}$$

which is a conic bundle over $\mathbb{C}^* \times \mathbb{C}^*$, where the conic fiber degenerates to two lines over the family of Riemann surfaces with punctures

$$\mathbb{S}(z) := \{H(x, y; z^\alpha) = 0\} \subset \mathbb{C}^* \times \mathbb{C}^* ,\tag{5.1.8}$$

parameterized by the complex parameters z^α. To establish that W is a non-compact Calabi-Yau manifold note that

$$\Omega = \frac{dH dx dy}{H x y}\tag{5.1.9}$$

is a regularizable no-where vanishing holomorphic volume form on W. Its periods are regularizable in the sense that H, y can be integrated out to yield a meromorphic one-form on $\mathbb{S}$

$$\lambda = \frac{\log(y) \mathrm{d}x}{x} ,\tag{5.1.10}$$

whose periods clearly exist. They are annihilated by the linear differential operators

$$D_\alpha = \prod_{Q_i^\alpha > 0} \partial_{x_i}^{Q_i^\alpha} - \prod_{Q_i^\alpha < 0} \partial_{x_i}^{-Q_i^\alpha} .\tag{5.1.11}$$

---

[5] The x_i here should not be identified with the x_i, which describe the A model in the previous section.

The redundancy in the parameterization of the complex structure is removed using the relations (5.1.5) and the scaling relation (5.1.4). To do that it is convenient to write the differential operator (5.1.11) in terms of logarithmic derivatives $\theta_i := x_i \partial_{x_i}$ and transform to logarithmic derivatives $\Theta_\alpha := z_\alpha \partial_{z_\alpha}$ using $\theta_i = Q_i^\alpha \Theta_\alpha$.
The solutions to (5.1.11) are constructed by the Frobenius method [70], i.e. defining

$$w_0(\underline{z}, \underline{\rho}) = \sum_{\underline{n}^\alpha} \frac{1}{\prod_i \Gamma[Q_i^\alpha(n^\alpha + \rho^\alpha) + 1]}((-1)^{Q_0^\alpha} z^\alpha)^{n^\alpha}, \qquad (5.1.12)$$

then

$$X^0 = w_0(\underline{z}, \underline{0}) = 1, \qquad t^\alpha = \frac{\partial}{2\pi i \partial \rho^\alpha} w_0(\underline{z}, \underline{\rho})|_{\underline{\rho}=0} \qquad (5.1.13)$$

are solutions. Note that the flat coordinates t^α approximate $t^\alpha \sim \log(z^\alpha)$ in the limit $z^\alpha \to 0$. Higher derivatives

$$X^{(\alpha_{i_1}\ldots\alpha_{i_n})} = \frac{1}{(2\pi i)^n} \frac{\partial}{\partial \rho^{\alpha_{i_1}}} \cdots \frac{\partial}{\partial \rho^{\alpha_{i_n}}} w_0(\underline{z}, \underline{\rho})|_{\underline{\rho}=0} \qquad (5.1.14)$$

also obey the recursion imposed by (5.1.11), *i.e.* they fulfill (5.1.11) up to finitely many terms. However, a unique, up to addition of previous solutions, linear combinations of the $X^{\alpha_{i_1}\ldots\alpha_{i_2}}$ is actually the last solution of the Picard-Fuchs system. This solution encodes the genus zero Gromov-Witten invariants. It is a derivative of the holomorphic prepotential $\mathcal{F}_0$ and the triple intersection $C_{ijk} = \partial_{t_i} \partial_{t_j} \partial_{t_k} \mathcal{F}_0$ can be constructed from it, see the examples for more details. We will turn to generating functions for the higher genus amplitudes in the next section.

5.2 Direct Integration in local Calabi-Yau geometries

Let us rephrase here the direct integration procedure explained in section 3.7 in the context of local Calabi-Yau manifolds. Recall that the key input for the direct integration procedure is the special geometry integration condition

$$\bar{\partial}_{\bar{\imath}} \Gamma_{ij}^k = \delta_i^k G_{j\bar{\imath}} + \delta_j^k G_{i\bar{\imath}} - C_{ijl} \bar{C}_{\bar{\imath}}^{kl} \ . \qquad (5.2.1)$$

Here C_{ijl} are the holomorphic Yukawa couplings which transform as $\text{Sym}^3(T\mathcal{M}) \otimes \mathcal{L}^{-2}$ and $\bar{C}_{\bar{\imath}}^{kl} = e^{2K} G^{k\bar{k}} G^{l\bar{l}} \bar{C}_{\bar{\imath}\bar{k}\bar{l}}$.
(5.2.1) implies that the propagator S^{ij}, which is defined by $\bar{\partial}_{\bar{k}} S^{ij} = \bar{C}_{\bar{k}}^{ij}$, can be solved from the integrated version of (5.2.1)

$$\Gamma_{ij}^k = \delta_i^k \partial_j K + \delta_j^k \partial_i K - C_{ijl} S^{kl} + \tilde{f}_{ij}^k \ , \qquad (5.2.2)$$

up to the holomorphic ambiguity $\tilde{f}_{ij}^k$. Taking the anti holomorphic derivative, using (5.2.1) and $\partial_{\bar{\jmath}} S^k = S_{\bar{\jmath}}^k$ it follows that

$$\bar{\partial}_{\bar{k}}(D_i S^{kl}) = \bar{\partial}_{\bar{k}}(\delta_i^k S^l + \delta_i^l S^k - C_{inm} S^{km} S^{ln}) \ , \qquad (5.2.3)$$

and so
$$D_i S^{kl} = \delta_i^k S^l + \delta_i^l S^k - C_{inm} S^{km} S^{ln} + f_i^{kl}. \tag{5.2.4}$$

In the local case one has the following simplifications. The Kähler connection in D_i becomes trivial due to the relation $\exp(K) \sim X^0 \to 1$ in the holomorphic limit in the non-compact models [76], and the S^l, (as well as the S, see 3.7.5) vanish, *i.e.* the above equation becomes simply
$$D_i S^{kl} = -C_{inm} S^{km} S^{ln} + f_i^{kl}. \tag{5.2.5}$$

Also, the Kähler connection $\partial_j K$ in (5.2.2) drops out, so the S^{ij} are solved from
$$\Gamma_{ij}^k = -C_{ijl} S^{kl} + \tilde{f}_{ij}^k. \tag{5.2.6}$$

Note that this is an over-determined system in the multi moduli case which requires a suitable choice of the ambiguity $\tilde{f}_{ij}^k$. This choice is simplified by the fact [77] that $\partial_i F^{(1)} 1$ can be expressed through the propagator as
$$\partial_i \mathcal{F}^{(1)} = \frac{1}{2} C_{ijk} S^{jk} + A_i, \tag{5.2.7}$$

with an ambiguity A_i, which can be determined by the ansatz $A_i = \partial_i (\tilde{a}_j \log \Delta_j + \tilde{b}_j \log z_j)$.

Once the S^{ij} are obtained and the ambiguities in (5.2.5,5.2.6) have been fixed, the direct integration of (3.5.35) is quite simple. Everything on the right hand side of the holomorphic anomaly equation (3.5.35) can be rewritten in terms of the generators S^{ij} and holomorphic functions. This way equation (3.7.8) transforms to
$$\frac{\partial \mathcal{F}^{(g)}}{\partial S^{jk}} = \frac{1}{2} (D_j \partial_k F_{g-1} + \sum_{r=1}^{g-1} \partial_j F_{g-r} \partial_k F_r). \tag{5.2.8}$$

This equation can easily be integrated w.r.t. S^{ij} and it can be shown that $\mathcal{F}^{(g)}$ is a polynomial in S^{jk} of degree $3g - 3$.

This leaves us with the holomorphic ambiguity. In the case of the local Calabi-Yau manifolds we are discussing here, the discriminant locus contains apart from the components $z_1 = 0$ and $z_2 = 0$ only one conifold divisor given by $\Delta = 0$. The gap condition (3.7.27) at the conifold divisor and regularity of the amplitudes everywhere else in the moduli space then leads to the following ansatz for the ambiguity
$$f_g = \frac{A_g}{\Delta^{2g-2}}, \tag{5.2.9}$$

where A_g is a polynomial of degree $(2g-2) \cdot \Delta$ in the z_i.

5.3 $\mathbb{K}_{\mathbb{P}^1 \times \mathbb{P}^1} = \mathcal{O}(-2,-2) \to \mathbb{P}^1 \times \mathbb{P}^1$

We are considering the non-compact Calabi-Yau geometry $\mathcal{O}(-2,-2) \to \mathbb{P}^1 \times \mathbb{P}^1$, i.e. the canonical line bundle over the Hirzebruch surface $\mathbb{F}_0 = \mathbb{P}^1 \times \mathbb{P}^1$. This local model can be obtained from the compact elliptic fibration over $\mathbb{F}_0$ with fiber $X_6(1,2,3)$. The three complexified Kähler volumes have the corresponding Mori cone generators $(-6;3,2,1,0,0,0,0)$, $(0;0,0,-2,1,0,1,0)$, $(0;0,0,-2,0,1,0,1)$. Roughly, in the local limit the volume of the elliptic fiber is send to infinity. The B-model mirror description of the local geometry is encoded in a Riemann surface with a meromorphic differential as pointed out before. According to [78] and using the above mentioned charge vectors, one can derive a Picard-Fuchs system governing the periods of the global mirror geometry. They are given by

$$\begin{aligned}\mathcal{D}_1 &= \Theta_1(\Theta_1 - 2\Theta_2 - 2\Theta_3) - 18z_1(1+6\Theta_1)(5+6\Theta_1) \\ \mathcal{D}_2 &= \Theta_2^2 + z_2(1 - \Theta_1 + 2\Theta_2 + 2\Theta_3)(\Theta_1 - 2\Theta_2 - 2\Theta_3) \\ \mathcal{D}_3 &= \Theta_3^2 + z_3(1 - \Theta_1 + 2\Theta_2 + 2\Theta_3)(\Theta_1 - 2\Theta_2 - 2\Theta_3),\end{aligned} \quad (5.3.1)$$

where we denote the logarithmic derivative by $\Theta_i = z_i \frac{\partial}{\partial z_i}$. z_1 is the complex structure parameter dual to the Kähler parameter of the elliptic fiber t_F. The local limit is obtained by sending this parameter to zero, $z_1 \to 0$.

Now let us turn to the non-compact geometry. The toric data of local $\mathbb{F}_0$ is summarized in the following matrix, V denoting the vectors which span the fan and Q denoting the charge vectors.

$$(V|Q) = \begin{pmatrix} 0 & 0 & 1 & -2 & -2 \\ 1 & 0 & 1 & 1 & 0 \\ 0 & -1 & 1 & 0 & 1 \\ -1 & 0 & 1 & 1 & 0 \\ 0 & 1 & 1 & 0 & 1 \end{pmatrix} \quad (5.3.2)$$

From there we conclude the following quantities as was explained in section 5.1.2. $C_{ijk}^{(0)}$ denote the classical triple intersection numbers. They, as well as $\int_M c_2 J_i$, were computed using toric geometry.

a) $Q^1 = (-2,1,0,1,0)$, $Q^2 = (-2,0,1,0,1)$
b) $Z = \{x_1 = x_3 = 0\} \cup \{x_2 = x_4 = 0\}$
c) $M = (\mathbb{C}^5[x_0,\cdots,x_4] \setminus Z)/(\mathbb{C}^*)^2$
d) $H(x,y) = y^2 - x^3 - (1 - 4z_1 - 4z_2)x^2 - 16z_1z_2 x$
e) $\mathcal{D}_1 = \Theta_1^2 - 2z_1(\Theta_1 + \Theta_2)(1 + 2\Theta_1 + 2\Theta_2)$
$\mathcal{D}_2 = \Theta_2^2 - 2z_2(\Theta_1 + \Theta_2)(1 + 2\Theta_1 + 2\Theta_2)$
$\Delta = 1 - 8(z_1 + z_2) + 16(z_1 - z_2)^2$
f) $C_{111}^{(0)} = \frac{1}{4}$, $C_{112}^{(0)} = -\frac{1}{4}$, $C_{122}^{(0)} = -\frac{1}{4}$, $C_{222}^{(0)} = \frac{1}{4}$
g) $\int_M c_2 J_1 = \int_M c_2 J_2 = -1$.

$\quad (5.3.3)$

$H(x,y) = 0$ defines a family of elliptic curves $\Sigma(z_1, z_2)$ whose j-function is given by

$$j(z_1, z_2) = \frac{((1 - 4z_1 - 4z_2)^2 - 48z_1 z_2)^3}{z_1^2 z_2^2 (1 - 8(z_1+z_2) + 16(z_1-z_2)^2)}. \quad (5.3.4)$$

5.3.1 Review of the moduli space $\mathcal{M}$

The moduli space, $\mathcal{M}$, of the local Calabi-Yau $\mathcal{O}(-2,-2) \to \mathbb{P}^1 \times \mathbb{P}^1$ is spanned by two Kähler moduli controlling the sizes of the two $\mathbb{P}^1$'s. The B-model mirror description of this geometry can be expressed through a Riemann surface together with a meromorphic differential. The meromorphic differential is the reduction of the holomorphic three-form of the mirror geometry to a one-form living on a Riemann surface as described in section 5.1.2. In our particular case we get a genus one Riemann surface with two non-trivial cycles. Apart from these the meromorphic differential has a residue arising from integration over a certain trivial cycle. Together these periods parameterize the two complex structure moduli which are mirror to the two Kähler moduli of the original model. The period integrals satisfy two linear differential equations of order two, given by the Picard-Fuchs operators. It is well known that these periods can at worse have logarithmic singularities. The singular locus in the moduli space can be obtained by calculating the discriminant of the Picard-Fuchs system (5.3.3). This yields

$$z_1 z_2 \left(1 - 8(z_1 + z_2) + 16(z_1 - z_2)^2\right) =: z_1 z_2 \Delta = 0. \tag{5.3.5}$$

One sees that the singular locus splits into three irreducible components given by the divisors $z_1 = 0$, $z_2 = 0$ and $\Delta = 0$. The moduli z_1, z_2 are compactified to $\mathbb{P}^2$. At the large complex structure point $L_1 \cap L_2$, two of the periods, $t_1 = \log(z_1) + \mathcal{O}(z)$ and $t_2 = \log(z_2) + \mathcal{O}(z)$, give the classical large Kähler volumes of the two $\mathbb{P}^1$. As C touches L_1 at $z_2 = \frac{1}{4}$, L_2 at $z_1 = \frac{1}{4}$ and I at $u = \frac{z_1}{z_1 + z_2} = \frac{1}{2}$ and all intersections are with contact order two, the Picard-Fuchs system cannot be solved around these points in moduli space. Therefore, the moduli space has to be blown up around these points so that all divisors have normal crossings. This is done by introducing two new divisors at each of these points which is depicted in figure 5.1.

More details about this moduli space can be found in [77]. For us the most relevant points are $I \cap F$ which is a $\mathbb{Z}_2$ orbifold point admitting a matrix model expansion, and the conifold locus C, relevant for fixing the holomorphic ambiguity of the free energy functions.

5.3.2 Solving the topological string on local $\mathbb{F}_0$ at large radius

By the method of Frobenius one can calculate the periods eliminated by the Picard-Fuchs system. As the charge vectors are chosen such that they span the Mori cone, the periods are calculated at the large radius point of the moduli space $\mathcal{M}(M)$. It is well known that the regular solution for this local model is simply $\omega_0(\underline{z}, 0) = 1$. Therefore the mirror map is equal to the single logarithmic solution and given by

$$2\pi i T_1(z_1, z_2) = \log z_1 + 2(z_1 + z_2) + 3(z_1^2 + 4z_1 z_2 + z_2^2) + \tfrac{20}{3}(z_1^3 + 9z_1^2 z_2 + 9z_1 z_2^2 + z_2^3) + \mathcal{O}(z^4)$$
$$2\pi i T_2(z_1, z_2) = \log z_2 + 2(z_1 + z_2) + 3(z_1^2 + 4z_1 z_2 + z_2^2) + \tfrac{20}{3}(z_1^3 + 9z_1^2 z_2 + 9z_1 z_2^2 + z_2^3) + \mathcal{O}(z^4). \tag{5.3.6}$$

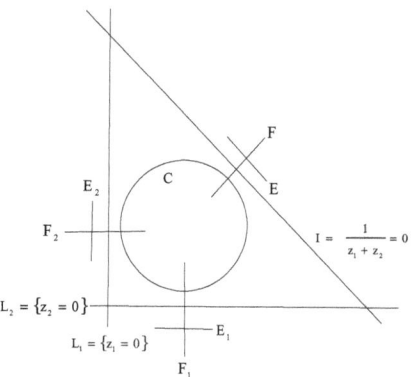

Figure 5.1: Resolved moduli space of $\mathbb{F}_0$.

By inverting the above series we arrive at $(Q_i = e^{2\pi i T_i})$

$$z_1(Q_1, Q_2) = Q_1 - 2(Q_1^2 + Q_1Q_2) + 3(Q_1^3 + Q_1Q_2^2) - 4(Q_1^4 + Q_1^3Q_2 + Q_1^2Q_2^2 + Q_1Q_2^3) + \mathcal{O}(Q^5)$$
$$z_2(Q_1, Q_2) = Q_2 - 2(Q_1Q_2 + Q_2^2) + 3(Q_1^2Q_2 + Q_2^3) - 4(Q_1^3Q_2 + Q_1^2Q_2^2 + Q_1Q_2^3 + Q_2^4) + \mathcal{O}(Q^5).$$
(5.3.7)

We observe that the following combination does not receive any instanton corrections which can be easily derived from the Picard-Fuchs system

$$\frac{z_1}{z_2} = \frac{Q_1}{Q_2} = e^{2\pi i(T_1 - T_2)} =: Q_1^x,$$
(5.3.8)

or in other words, the mirror map can be brought in trigonal form by means of the coordinate choice, $x_1 = \frac{z_1}{z_2}$ and $x_2 = z_2$, as well as $Q_2^x = Q_2$. We have

$$x_1(Q_1^x, Q_2^x) = Q_1^x,$$
$$x_2(Q_1^x, Q_2^x) = Q_2^x - 2Q_2^{x2} + Q_1^xQ_2^{x2} + 3Q_2^{x3} + \mathcal{O}(Q^4).$$
(5.3.9)

The next step is to determine the Yukawa couplings. Four independent combinations are

$$C_{111} = \frac{(1 - 4z_2)^2 - 16z_1(1 + z_1)}{4z_1^3\Delta}, \quad C_{112} = \frac{16z_1^2 - (1 - 4z_1)^2}{4z_1^2 z_2 \Delta},$$
$$C_{122} = \frac{16z_2^2 - (1 - 4z_2)^2}{4z_1 z_2^2 \Delta}, \quad C_{222} = \frac{(1 - 4z_1)^2 - 16z_1(1 + z_2)}{4z_2^3 \Delta}.$$
(5.3.10)

The numerator is fixed by the help of the known classical triple intersection numbers as well as the genus zero GV invariants, whereas the denominator is fixed by the Picard-Fuchs system. Note, that the Yukawa couplings are of the well-known structure, i.e. a rational

5.3. $\mathbb{K}_{\mathbb{P}^1 \times \mathbb{P}^1} = \mathcal{O}(-2,-2) \to \mathbb{P}^1 \times \mathbb{P}^1$

function in the z_i's multiplied by the inverse of the discriminant. Here we note, that in local models the choice of the classical data is crucial for the success of direct integration. This is due to the fact, that one can obtain the right GV invariants for different choices of $C^{(0)}$ and $\int c_2 J$. However, if one does not use consistent data, higher genus calculations become wrong or even impossible. In contrast, the dependence on some Euler number drops out completely, as it does not effect the GV invariants. In this work we simply set χ to zero.

Using the ansatz (3.7.23) for the free energy function of genus one and the classical data $\int c_2 J_i$ as well as the known genus one GV invariants we are able to fix the holomorphic ambiguity at genus one, f_1. The result as well as the expansion at large radius in the holomorphic limit $\overline{T} \to 0$ reads as follows

$$\mathcal{F}^{(1)} = \log\left(\Delta^{-\frac{1}{12}}(z_1 z_2)^{-\frac{13}{24}}(\det(G_{i\bar{j}}))^{-\frac{1}{2}}\right),$$

$$F^1(T_1, T_2) = -\frac{1}{24}\log(Q_1 Q_2) - \frac{1}{6}(Q_1 + Q_2) - \frac{1}{12}(Q_1^2 + 4Q_1 Q_2 + Q_2^2) + \mathcal{O}(Q^3). \tag{5.3.11}$$

In order to perform the method of direct integration, we have to calculate the propagator and express all quantities which carry non-holomorphic information through our propagators. As a first step the holomorphic ambiguity, $\tilde{f}$, in (5.2.6) can be fixed by the choice

$$\tilde{f}^1_{11} = -\frac{1}{z_1}, \quad \tilde{f}^1_{12} = -\frac{1}{4z_2}, \quad \tilde{f}^1_{22} = 0,$$
$$\tilde{f}^2_{11} = 0, \quad \tilde{f}^2_{12} = -\frac{1}{4z_1}, \quad \tilde{f}^2_{22} = -\frac{1}{z_2}, \tag{5.3.12}$$

where all other combinations follow by symmetry. We note that the propagator has only one independent component for we can write

$$S^{ij} = \begin{pmatrix} S(z_1, z_2) & \frac{z_2}{z_1} S(z_1, z_2) \\ \frac{z_2}{z_1} S(z_1, z_2) & \frac{z_2^2}{z_1^2} S(z_1, z_2) \end{pmatrix} \tag{5.3.13}$$

where $S(z_1, z_2) = \frac{1}{2}z_1^2 - 2z_1^3 - 2z_1^2 z_2 - 8z_1^3 z_2 - 32z_1^4 z_2 + \mathcal{O}(z^6)$. This is due to the fact, that the mirror geometry is solely determined by the elliptic curve $\Sigma(z_1, z_2)$, which has only one relevant elliptic parameter τ. The dependence on a second parameter is due to a non-vanishing residue of the meromorphic differential on $\Sigma(z_1, z_2)$.

Often it is convenient and also more natural to perform the calculations in the coordinates x_1, x_2, in which some Christoffel symbols are rational

$$\Gamma^1_{11} = \frac{1}{x_1}, \quad \Gamma^1_{12} = 0, \quad \Gamma^1_{22} = 0.$$

Noting, that from the tensorial transformation law of the propagator and the relation (5.2.6) the ambiguity of the propagator $\tilde{f}$ has to transform as $\tilde{f}^i_{jk}(x) = \frac{\partial x_i}{\partial z_l}\left(\frac{\partial^2 z_l}{\partial x_j \partial x_k}\right) +$

$\frac{\partial x_i}{\partial z_l} \frac{\partial z_m}{\partial x_j} \frac{\partial z_n}{\partial x_k} \tilde{f}^l_{mn}(x(z))$. We obtain

$$\tilde{f}^1_{11} = -\frac{1}{x_1}, \quad \tilde{f}^2_{12} = -\frac{1}{4x_1}, \quad \tilde{f}^2_{22} = -\frac{3}{2x_2}, \tag{5.3.14}$$

where all other combinations are either 0 or follow by symmetry. As $\Gamma^1_{ij} = -\tilde{f}^1_{ij}$ we observe that the propagator takes the following simple form $S^{11} = S^{12} = S^{21} = 0$ and $S^{22} = \frac{x_2^2}{2} - 2x_2^3 - 2x_1x_2^3 + \mathcal{O}(x^5)$.
In addition, we fix the holomorphic ambiguity of the covariant derivative of S^{ij}, (5.2.5), and obtain

$$f^{11}_1 = -\tfrac{1}{8}z_1(1+4z_1-4z_2), \; f^{12}_1 = -\tfrac{1}{8}z_2(1+4z_1-4z_2), \; f^{22}_1 = -\tfrac{z_2^2}{8z_1}(1+4z_1-4z_2),$$
$$f^{11}_2 = -\tfrac{z_1^2}{8z_2}(1+4z_2-4z_1), \; f^{12}_2 = -\tfrac{1}{8}z_1(1+4z_2-4z_1), \; f^{22}_2 = -\tfrac{1}{8}z_2(1+4z_2-4z_1), \tag{5.3.15}$$

where all other combinations follow by symmetry. Further we can express the covariant derivative of $\mathcal{F}(1)$ through the generator S (5.2.7) by

$$D_i\mathcal{F}^{(1)} = \frac{1}{2}C_{ijk}S^{jk} - \frac{1}{12}\Delta^{-1}\partial_i\Delta + \frac{7}{24z_i}. \tag{5.3.16}$$

Note, that in contrast to an one parameter model the holomorphic ambiguity $A_i = \partial_i(\tilde{a}_j \log \Delta_j + \tilde{b}_j \log z_j)$ in (5.3.16) cannot be set to zero. More generally, in the local models we are considering here the geometry of the B-model is encoded in a Riemann surface of genus one whose moduli space admits only one quasimodular form of weight 2, namely the second Eisenstein series. Therefore and from the discussions in the case of local $\mathbb{P}^2$ in the previous section we expect there to be a coordinate system in which the propagator is proportional to the second Eisenstein series. The relevant coordinate system is given by the x-coordinates in which it is allowed to set all but one component of the propagator to zero and subsequently one can use (5.2.7) and (3.7.23) to solve for this non-zero component. Now, in the multi-parameter case this gives, for each direction of the derivative of $\mathcal{F}^{(1)}$ w.r.t. z_i, $h^{2,1}$ equations on $\tilde{a}_j, \tilde{b}_j$. In this and the following example, we are lucky as these constraints fix the parameters completely. In addition one arrives at a series expansion for the non-vanishing component of S^{ij}. This can be used to fix all ambiguities in the model as rational functions of the z_i with poles only at the singular divisors of the Picard-Fuchs system.

Now, all input to perform direct integration is provided and applying this method we are able to determine $\mathcal{F}^{(g)}$ for genus g up to four. Using that local $\mathbb{F}_0$ has a discriminant with $\deg \Delta = 2$ and we can further reduce the number of coefficients in A_g due to symmetry in z_1 and z_2, one can easily calculate, that at genus g there are $(2g-1)^2$ unknowns in the holomorphic ambiguity. Therefore genus four corresponds to fixing 49 coefficients in the holomorphic ambiguity $f_g = \frac{A_g}{\Delta^{2g-2}}$. They are determined by the gap condition at the conifold locus and the known constant map contributions. We will further comment on this in the next section.

5.3. $\mathbb{K}_{\mathbb{P}^1\times\mathbb{P}^1} = \mathcal{O}(-2,-2) \to \mathbb{P}^1 \times \mathbb{P}^1$

Let's present at least the genus two results. The free energy is given by

$$\mathcal{F}^{(2)} = \frac{5}{24z_1^6\Delta^2}S^3 + \frac{-13 + 48z_1^2 + z_1(40 - 96z_2) + 40z_2 + 48z_2^2}{48z_1^4\Delta^2}S^2 +$$
$$\frac{384z_1^3 + z_1^2(80 - 384z_2) + (1 - 4z_2)^2(17 + 24z_2) - 16z_1(7 - 46z_2 + 24z_2^2)}{144z_1^2\Delta^2}S + f_2,$$
(5.3.17)

where the ambiguity $f_2 = \frac{A_2}{\Delta^2}$ is fixed by the following choice

$$A_2 = -\frac{1}{1440}(25 - 258z_1 + 696z_1^2 + 416z_1^3 - 2688z_1^4 - 258z_2 + 2768z_1z_2 - 6560z_1^2z_2 - 1536z_1^3z_2$$
$$+ 696z_2^2 - 6560z_1z_2^2 + 8448z_1^2z_2^2 + 416z_2^3 - 1536z_1z_2^3 - 2688z_2^4).$$
(5.3.18)

The solution around the conifold is described in the next section. The GV invariants can be found in the appendix D.1. They are in accord with [79] as far as they have been computed.

5.3.3 Solving the topological string on local $\mathbb{F}_0$ at the conifold locus

Our next task is to solve the Picard-Fuchs equations around the conifold locus. In order to do that we choose some convenient point on the locus and define variables which are good coordinates around this point. In our case we choose the point to be $z_1 = \frac{1}{16}, z_2 = \frac{1}{16}$. As one can easily check inserting these numbers into the discriminant yields zero. To find the right variables we have to be careful as their gradients at the relevant point must not be colinear. The following choice will do the job

$$z_{c,1} = 1 - \frac{z_1}{z_2}, \quad z_{c,2} = 1 - \frac{z_2}{\frac{1}{8} - z_1}.$$
(5.3.19)

We transform the Picard-Fuchs system to the above coordinates and find the following polynomial solutions

$$\omega_0^c = 1,$$
$$\omega_1^c = -\log(1 - z_{c,1}),$$
$$\omega_2^c = z_{c,2} + \frac{1}{16}(2z_{c,1}^2 + 8z_{c,1}z_{c,2} + 13z_{c,2}^2) + \mathcal{O}(z_c^3).$$
(5.3.20)

As mirror coordinates we take $t_{c,1} := \omega_1^c$ and $t_{c,2} := \omega_2^c$. Inverting these series gives the following mirror map

$$z_{c,1}(t_{c,1}, t_{c,2}) = 1 - e^{-t_{c,1}},$$
$$z_{c,2}(t_{c,1}, t_{c,2}) = t_{c,2} - \frac{1}{16}(t_{c,1}^2 + 8t_{c,1}t_{c,2} + 13t_{c,2}^2) + \mathcal{O}(t_c^5).$$
(5.3.21)

CHAPTER 5. LOCAL CALABI-YAU BACKGROUNDS

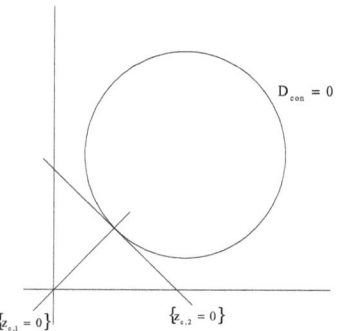

Figure 5.2: Conifold coordinates.

The divisor $\{z_{c,1} = 0\}$ is normal to the conifold locus at $(z_1, z_2) = \left(\frac{1}{16}, \frac{1}{16}\right) = p_{\text{con}}$ whereas $\{z_{c,2} = 0\}$ is tangential (see figure 5.2). Therefore $z_{c,1}$ parameterizes the tangential direction to the conifold locus at p_{con} in moduli space and $z_{c,2}$ the normal one. Hence we expect the flat mirror coordinate $t_{c,2}$ to be controlling the size of the shrinking cycle at p_{con}, thus $t_{c,2}$ should appear in inverse powers in the expansion of the free energies.

Transforming the Yukawa couplings, the Christoffel symbols and the holomorphic ambiguities $\tilde{f}$ to the conifold coordinates we obtain the propagator around this locus. In the choice of our coordinates (5.3.19) the propagator takes the following simple form $S^{11} = S^{12} = S^{21} = 0$ and

$$S^{22} = \frac{1}{2}t_{c,2} + \frac{1}{1536}(24 t_{c,1}^2 t_{c,2} + t_{c,2}^3) + \mathcal{O}(t_c^4).$$

Assuming the gap condition holds, we are able to fix all but one coefficients of the holomorphic ambiguity. Expanding the free energies at the large radius point in moduli space the constant map contribution fixes the last unknown, i.e. we observe that the gap condition yields at genus two 8 out of 9 unknowns, at genus three 24 out of 25 unknowns, etc. Our results up to genus four are given below (rescaling: $t_{c,2} \to 2t_{c,2}$)

$$F_c^2 = -\frac{1}{240 t_{c,2}^2} - \frac{1}{1152} + \frac{53 t_{c,2}}{122880} + \frac{t_{c,1}^2}{61440} - \frac{2221 t_{c,2}^2}{14745600} + \mathcal{O}(t_c^3)$$

$$F_c^3 = \frac{1}{1008 t_{c,2}^4} + \frac{23}{5806080} + \frac{407 t_{c,2}}{198180864} - \frac{t_{c,1}^2}{3096576} - \frac{258485 t_{c,2}^2}{49941577728} + \mathcal{O}(t_c^3) \quad (5.3.22)$$

$$F_c^4 = -\frac{1}{1440 t_{c,2}^6} - \frac{19}{278691840} + \frac{114773 t_{c,2}}{362387865600} + \mathcal{O}(t_c^2).$$

5.3.4 Solving the topological string on local $\mathbb{F}_0$ at the orbifold point

As we have noted already there exists an orbifold point in the moduli space $\mathcal{M}$ at which we can compare our results with the known matrix model expansions.
At this point we expand the periods in the local variables

$$z_{o,1} = 1 - \frac{z_1}{z_2}, \quad z_{o,2} = \frac{1}{\sqrt{z_2}\left(1 - \frac{z_1}{z_2}\right)}. \tag{5.3.23}$$

Transforming the Picard-Fuchs system to these coordinates and solving it, we obtain the following set of periods

$$\begin{aligned}
\omega_0^o &= 1, \\
\omega_1^o &= -\log(1 - z_{o,1}), \\
\omega_2^o &= z_{o,1} z_{o,2} + \frac{1}{4} z_{o,1}^2 z_{o,2} + \frac{9}{64} z_{o,1}^3 z_{o,2} + \mathcal{O}(z_o^5), \\
F_{\omega_2^o}^{(0)} &= \omega_2^o \log(z_{o,1}) + \frac{1}{2} z_{o,1}^2 z_{o,2} + \frac{21}{64} z_{o,1}^3 z_{o,2} + \mathcal{O}(z_o^5).
\end{aligned} \tag{5.3.24}$$

We define the mirror map to be given by the first two periods

$$t_{o,1} := \omega_1^o, \quad t_{o,2} := \omega_2^o, \tag{5.3.25}$$

and will express the B-model correlators in terms of these coordinates. In order to invert the mirror map and find the function $z_o(t_o)$, we have to consider the two series $\tilde{t}_{o,1} = t_{o,1} = z_{o,1} + 1 + \mathcal{O}(z_o^2)$ and $\tilde{t}_{o,2} = \frac{t_{o,2}}{t_{o,1}} = z_{o,2} + \mathcal{O}(z_o^2)$. Inverting these we obtain

$$\begin{aligned}
z_{o,1}(\tilde{t}_{o,1}) &= 1 - e^{-\tilde{t}_{o,1}}, \\
z_{o,2}(\tilde{t}_{o,1}, \tilde{t}_{o,2}) &= \tilde{t}_{o,2} + \frac{1}{4}\tilde{t}_{o,1}\tilde{t}_{o,2} + \frac{1}{192}\tilde{t}_{o,1}^2\tilde{t}_{o,2} - \frac{1}{256}\tilde{t}_{o,1}^3\tilde{t}_{o,2} + \mathcal{O}(\tilde{t}_o^5),
\end{aligned} \tag{5.3.26}$$

which together form the mirror map at the orbifold point in moduli space.
Transforming the Yukawa couplings, the Christoffel symbols and the holomorphic ambiguities $\tilde{f}$ to the orbifold coordinates we obtain the propagator around this locus. In the choice of our coordinates (5.3.23) the propagator takes the following simple form $S^{11} = S^{12} = S^{21} = 0$ and

$$S^{22} = \frac{1}{16}(t_{o,2}^2 - t_{o,1}^2) + \frac{1}{6144}(t_{o,1}^4 - 6t_{o,1}^2 t_{o,2}^2 + 5t_{o,2}^4) + \mathcal{O}(t_o^5).$$

In order to match the matrix model expansion one has to choose appropriate coordinates. As explained in [77] the right variables S_1, S_2 that match the 't Hooft parameters on the matrix model side are given by

$$S_1 = \frac{1}{4}(t_{o,1} + t_{o,2}), \quad S_2 = \frac{1}{4}(t_{o,1} - t_{o,2}). \tag{5.3.27}$$

In addition the overall normalization of the all genus partition function $\mathcal{F} = \sum_g g_s^{2g-2} \mathcal{F}^{(g)}$ has to be determined. By comparing to the matrix model one gets, that the string coupling on the topological side, g_s^{top}, is related to the coupling on the matrix model side, $\hat{g}_s$, by the identification $g_s^{\text{top}} = 2i\hat{g}_s$. Using these expressions we find

$$F_{orb}^2 = -\frac{1}{240}\left(\frac{1}{S_1^2} + \frac{1}{S_2^2}\right) + \frac{1}{360} - \frac{1}{57600}(S_1^2 + 60S_1S_2 + S_2^2) + \mathcal{O}(S^4)$$

$$F_{orb}^3 = \frac{1}{1008}\left(\frac{1}{S_1^4} + \frac{1}{S_2^4}\right) + \frac{1}{22680} + \frac{1}{34836480}(S_1^2 - 252S_1S_2 + S_2^2) + \mathcal{O}(S^4)$$

$$F_{orb}^4 = -\frac{1}{1440}\left(\frac{1}{S_1^6} + \frac{1}{S_2^6}\right) + \frac{1}{340200} - \frac{1}{82944000}(S_1^2 + 102S_1S_2 + S_2^2) + \mathcal{O}(S^4). \tag{5.3.28}$$

The genus two results are in accord with [77], genus three corrects the misprints in this article and genus four is a prediction on the matrix model.

5.3.5 Relation to the family of elliptic curves

At the beginning of this section we pointed out, that $H(x,y) = 0$ defines a family of elliptic curves $\Sigma(z_1, z_2)$ whose j-function is given by

$$j(z_1, z_2) = \frac{((1 - 4z_1 - 4z_2)^2 - 48z_1z_2)^3}{z_1^2 z_2^2 (1 - 8(z_1 + z_2) + 16(z_1 - z_2)^2)}. \tag{5.3.29}$$

Using the usual j-function description (B.1.11) one can establish a relation between the elliptic parameter $q = e^{2\pi i \tau}$ and the complex structure variables z_1 and z_2 which reads

$$q = z_1^2 z_2^2 + 16z_1^3 z_2^2 + 160z_1^4 z_2^2 + 16z_1^2 z_2^3 + 400z_1^3 z_2^3 + 160z_1^2 z_2^4 + \mathcal{O}(z^7). \tag{5.3.30}$$

We observe that

$$\tau = 4\partial_{t_{x,2}} \partial_{t_{x,2}} F^0, \quad \partial_{t_{x,2}} \tau = -4C_{t_{x,2} t_{x,2} t_{x,2}}, \tag{5.3.31}$$

where $t_{x,i}$ is obtained from $Q_i^x = e^{2\pi i t_{x,i}}$, which hints at that the not instanton corrected parameter x_1 or Q_1^x, respectively, is merely an auxiliary parameter.
[80] work with an isogenous description of $\tilde{\Sigma}(z_1, z_2)$. They use the Segre embedding of $\mathbb{P}^1 \times \mathbb{P}^1$ into $\mathbb{P}^3$ given by the map

$$([x_0 : x_1], [x_0' : x_1']) \mapsto [X_0 : X_1 : X_2 : X_3] = [x_0 x_0', x_1 x_0', x_0 x_1', x_1 x_1'], \tag{5.3.32}$$

where $[x_0 : x_1]$ and $[x_0' : x_1']$ are homogeneous coordinates of the $\mathbb{P}^1$'s and $X_0, \ldots, X_3$ are homogeneous coordinates of $\mathbb{P}^3$. Then $\tilde{\Sigma}(\tilde{z}_1, \tilde{z}_2)$ is given by the complete intersection of $\mathbb{P}^1 \times \mathbb{P}^1$, defined by $X_0 X_3 - X_1 X_2$, with the hypersurface given by $X_0^2 + \tilde{z}_1 X_1^2 + X_2^2 + \tilde{z}_2 X_3^2 + X_0 X_3$. Its j-function reads

$$\tilde{j}(\tilde{z}_1, \tilde{z}_2) = \frac{((1 - 4\tilde{z}_1 - 4\tilde{z}_2)^2 + 192\tilde{z}_1 \tilde{z}_2)^3}{\tilde{z}_1 \tilde{z}_2 (1 - 8(\tilde{z}_1 + \tilde{z}_2) + 16(\tilde{z}_1 - \tilde{z}_2)^2)^2}. \tag{5.3.33}$$

Defining $\tilde{q} = e^{2\pi i \tilde{\tau}}$ we can calculate that $\tilde{\tau} = \partial_{t_{x,2}}\partial_{t_{x,2}}F^0$, i.e. their modular parameters are related by a simple rescaling by a factor of 4

$$\tau = 4\tilde{\tau}. \tag{5.3.34}$$

This transfers to a rescaling of the periods of the elliptic curve.
With this input it is possible to write the full non-holomorphic $\mathcal{F}^{(1)}$ as

$$\mathcal{F}^{(1)} = -\log\sqrt{\tilde{\tau}_2}\eta(\tilde{\tau})\bar{\eta}(\bar{\tilde{\tau}}). \tag{5.3.35}$$

5.4 $\mathbb{K}_{\mathbb{F}_1} = \mathcal{O}(-2,-3) \to \mathbb{F}_1$

We are considering the non-compact Calabi-Yau geometry $\mathcal{O}(-2,-3) \to \mathbb{F}_1$, i.e. the canonical line bundle over the Hirzebruch surface $\mathbb{F}_1 = \mathbb{BP}_1^2$, where $\mathbb{BP}_1^2$ denotes the first del Pezzo surface, i.e. $\mathbb{P}^2$ with one blow up. This local model can be obtained again from the compact elliptic fibration over $\mathbb{F}_1$ with fiber $X_6(1,2,3)$. The three complexified Kähler volumes have the corresponding Mori cone generators $(-6;3,2,1,0,0,0,0)$, $(0;0,0,-1,1,-1,1,0)$, $(0;0,0,-2,0,1,0,1)$. A Picard-Fuchs system governing the periods of the global mirror geometry is given by

$$\begin{aligned}
\mathcal{D}_1 &= \Theta_1(\Theta_1 - 2\Theta_2 - \Theta_3) - 18z_1(1+6\Theta_1)(5+6\Theta_1) \\
\mathcal{D}_2 &= \Theta_2(\Theta_2 - \Theta_3) - z_2(-1+\Theta_1 - 2\Theta_2 - \Theta_3)(\Theta_1 - 2\Theta_2 - \Theta_3) \\
\mathcal{D}_3 &= \Theta_3^2 - z_3(\Theta_1 - 2\Theta_2 - \Theta_3)(\Theta_2 - \Theta_3).
\end{aligned} \tag{5.4.1}$$

Now let us turn to the non-compact geometry. The toric data of local $\mathbb{F}_1$ is summarized in the following matrix

$$(V|Q) = \begin{pmatrix} 0 & 0 & 1 & -2 & -1 \\ 1 & 0 & 1 & 1 & 0 \\ -1 & -1 & 1 & 0 & 1 \\ -1 & 0 & 1 & 1 & -1 \\ 0 & 1 & 1 & 0 & 1 \end{pmatrix}. \tag{5.4.2}$$

From there we conclude the following quantities[6]

a) $Q^1 = (-2,1,0,1,0)$, $Q^2 = (-1,0,1,-1,1)$
b) $Z = \{x_1 = x_3 = 0\} \cup \{x_2 = x_4 = 0\}$
c) $M = (\mathbb{C}^5[x_0,\cdots,x_4] \setminus Z)/(\mathbb{C}^*)^2$
d) $H(x,y) = y^2 - x^3 - (1-4z_1)x^2 + 8z_1z_2x - 16z_1^2z_2^2$
e) $\mathcal{D}_1 = \Theta_1(\Theta_1 - \Theta_2) - z_1(2\Theta_1 + \Theta_2)(1+2\Theta_1 + 2\Theta_2)$
   $\mathcal{D}_2 = \Theta_2^2 - z_2(\Theta_2 - \Theta_1)(2\Theta_1 + \Theta_2)$ \hfill (5.4.3)
   $\Delta = (1-4z_1)^2 - z_2(1 - 36z_1 + 27z_1z_2)$
f) $C_{111}^{(0)} = -\frac{1}{3}$, $C_{112}^{(0)} = -\frac{1}{3}$, $C_{122}^{(0)} = -\frac{1}{3}$, $C_{222}^{(0)} = \frac{2}{3}$
g) $\int_M c_2 J_1 = -2$, $\int_M c_2 J_2 = 0$.

---

[6] Using toric geometry it is only possible to determine an one-parameter family of classical intersection numbers $C_{ijk}^{(0)}$, resulting in an one-parameter family for $\int_M c_2 J_i$. Their correct values are fixed by a limiting procedure of local $\mathbb{F}_1 = \mathbb{BP}_1^2$ to local $\mathbb{P}^2$ which is described below.

$H(x,y) = 0$ defines a family of elliptic curves $\tilde{\Sigma}(z_1, z_2)$ whose j-function is given by

$$j(z_1, z_2) = \frac{((1-4z_1)^2 + 24z_1z_2)^3}{z_1^3 z_2^2((1-4z_1)^2 - z_2(1 - 36z_1 + 27z_1z_2))}. \quad (5.4.4)$$

5.4.1 Solving the topological string on local $\mathbb{F}_1$ at large radius

The mirror map at the point of large radius is given by

$$\begin{aligned} 2\pi i T_1(z_1, z_2) &= \log z_1 + 2z_1 + 3z_1^2 - 4z_1 z_2 + \tfrac{20}{3} z_1^3 + 24 z_1^2 z_2 + \mathcal{O}(z^4) \\ 2\pi i T_2(z_1, z_2) &= \log z_2 + z_1 + \tfrac{3}{2} z_1^2 - 2z_1 z_2 + \tfrac{10}{3} z_1^3 + -12 z_1^2 z_2 + \mathcal{O}(z^4). \end{aligned} \quad (5.4.5)$$

Inverting the series we obtain for $Q_i = e^{2\pi i T_i}$

$$\begin{aligned} z_1(Q_1, Q_2) &= Q_1 - 2Q_1^2 + 3Q_1^3 + 4Q_1^2 Q_2 - 4(Q_1^4 + Q_1^3 Q_2) + \mathcal{O}(Q^5) \\ z_2(Q_1, Q_2) &= Q_2 - Q_1 Q_2 + Q_1^2 Q_2 + 2Q_1 Q_2^2 - Q_1^3 Q_2 + \mathcal{O}(Q^5). \end{aligned} \quad (5.4.6)$$

Now, one realizes again that there is a relation between the Q coordinates:

$$\frac{Q_1}{Q_2^2} = \frac{z_1}{z_2^2} = e^{2\pi i (T_1 - 2T_2)} =: Q_1^x. \quad (5.4.7)$$

Defining further $Q_2^x := Q_2$ and $x_1 = \frac{z_1}{z_2^2}$ as well as $x_2 = z_2$ one finds that

$$\begin{aligned} x_1(Q_1^x, Q_2^x) &= Q_1^x, \\ x_2(Q_1^x, Q_2^x) &= Q_2^x - Q_1^x Q_2^{x3} + 2Q_1^x Q_2^{x4} + \mathcal{O}(Q^6). \end{aligned} \quad (5.4.8)$$

The Yukawa couplings can be fixed through the relation $\partial_{T_i} \partial_{T_j} \partial_{T_k} F^0 = C_{T_i T_j T_k}$ and the known genus zero GV invariants up to a dependence on one unfixed parameter. This unfixed parameter can be determined by the fact that there exists a limit of local $\mathbb{F}_1$ to local $\mathbb{P}^2$, as $\mathbb{F}_1 = B\mathbb{P}_1^2$. This blow-down limit turns out to be

$$z_1 \to 0, \text{ with } z_1 z_2 = z \text{ fixed.}$$

We obtain the following Yukawa couplings

$$C_{111} = \frac{-1 - 4z_1^2 + z_2 - z_1(7 - 6z_2)}{3z_1^3 \Delta}, \quad C_{112} = \frac{-1 + 8z_1^2 + z_2 + z_1(2 - 3z_2)}{3z_1^2 z_2 \Delta},$$
$$C_{122} = \frac{z_2(1 - 12z_1) - (1 - 4z_1)^2}{3z_1 z_2^2 \Delta}, \quad C_{222} = \frac{2(1 - 4z_1)^2 + z_2(1 - 60z_1)}{3z_2^3 \Delta}. \quad (5.4.9)$$

The next step is to determine the propagators of local $\mathbb{F}_1$. This is best done in x coordinates, where one finds again that some Christoffel symbols are either trivial or have a rational form

$$\Gamma_{11}^1 = -\frac{1}{x_1}, \quad \Gamma_{12}^1 = 0, \quad \Gamma_{22}^1 = 0. \quad (5.4.10)$$

5.4. $\mathbb{K}_{\mathbb{F}_1} = \mathcal{O}(-2,-3) \to \mathbb{F}_1$

Choosing $\tilde{f}_{11}^1 = -\frac{1}{x_1}$, $\tilde{f}_{12}^1 = 0$, $\tilde{f}_{21}^1 = 0$, $\tilde{f}_{22}^1 = 0$, one finds from (5.2.6) that S^{11}, S^{12} are immediately zero. Demanding symmetry we are able to fix all ambiguities $\tilde{f}_{jk}^i$ by the choice

$$\tilde{f}_{11}^1 = -\frac{1}{x_1}, \quad \tilde{f}_{11}^2 = -\frac{x_2}{12x_1^2\Delta_x}(1 - x_2 - 12x_1x_2^2 + 49x_1x_2^3 - 36x_1x_2^4 + 32x_1^2x_2^4 - 12x_1^2x_2^5),$$

$$\tilde{f}_{12}^2 = -\frac{1}{12x_1\Delta_x}(3 - 3x_2 - 32x_1x_2^2 + 144x_1x_2^3 - 108x_1x_2^4 + 80x_1^2x_2^4),$$

$$\tilde{f}_{22}^2 = -\frac{1}{12x_2\Delta_x}(20 - 21x_2 - 176x_1x_2^2 + 828x_1x_2^3 - 648x_1x_2^4 + 384x_1^2x_2^4),$$

(5.4.11)

where Δ_x denotes the discriminant in x coordinates and all other combinations of $\tilde{f}_{jk}^i$ are either zero or follow by symmetry. This singles out one non-vanishing propagator only, given by $S^{22}(x_1, x_2) = \frac{x_2^2}{12} - \frac{1}{3}x_1x_2^4 + x_1x_2^5 + 4x_1^2x_2^7 + \mathcal{O}(x^{10})$. After tensor transforming to z coordinates we obtain

$$S^{ij} = \begin{pmatrix} S(z_1, z_2) & \frac{z_2}{2z_1}S(z_1, z_2) \\ \frac{z_2}{2z_1}S(z_1, z_2) & \frac{z_2^2}{4z_1^2}S(z_1, z_2) \end{pmatrix}, \qquad (5.4.12)$$

where $S(z_1, z_2) = \frac{z_2^2}{3} - \frac{4z_1^3}{3} + 4z_1^3z_2 + 16z_1^4z_2 + \mathcal{O}(z^6)$. This again has a similar form as in the case of local $\mathbb{F}_0$.

In addition, we fix the holomorphic ambiguity of the covariant derivative of S^{ij}, (5.2.5), and obtain, that in x coordinates there are two non-zero contributions only, given by

$$f_1^{22} = -\frac{x_2^2}{144x_1\Delta_x}(3 - 3x_2 + 4x_1x_2^2)(1 - 8x_1x_2^2 + 24x_1x_2^3 + 16x_1^2x_2^4),$$

$$f_2^{22} = -\frac{x_2}{144\Delta_x}(8 - 9x_2)(1 - 8x_1x_2^2 + 24x_1x_2^3 + 16x_1^2x_2^4).$$

(5.4.13)

The f_i^{jk} in z coordinates are again obtained after tensor transformation.

Further we can express the covariant derivative of $\mathcal{F}^{1)}$ through the generator S by

$$D_i\mathcal{F}^{(1)} = \frac{1}{2}C_{ijk}S^{jk} + A_i. \qquad (5.4.14)$$

As the free energy function of genus one is given by

$$\mathcal{F}^{(1)} = \log\left(\Delta^{-\frac{1}{12}}z_1^{-\frac{7}{12}}z_2^{-\frac{1}{2}}\det(G_{i\bar{j}}))^{-\frac{1}{2}}\right),$$

$$F^1(T_1, T_2) = -\frac{1}{12}\log(Q_1) - \frac{1}{12}(2Q_1 + Q_2) - \frac{1}{24}(2Q_1^2 + 6Q_1Q_2 + Q_2^2) + \mathcal{O}(Q^3),$$

(5.4.15)

we find that $A_i = \partial_i A$ and

$$A = -\frac{1}{24}\log\Delta + \frac{1}{24}\log z_1 + \frac{1}{12}\log z_2. \qquad (5.4.16)$$

Now, we are prepared to perform the direct integration procedure. Demanding the gap at the conifold and using further the known constant map contributions we are able to fix the ambiguities up to genus three. In this more general two parameter model with one discriminant component of degree three the number of coefficients in A_g is

$$\binom{(2g-2)\deg\Delta+2}{2} = 10 - 27g + 18g^2, \qquad (5.4.17)$$

i.e. at genus three we have to fix 91 coefficients in the holomorphic ambiguity. The invariants can be found in the appendix D.1. The solutions around the conifold locus are described in the next section.

5.4.2 Solving the topological string on local $\mathbb{F}_1$ at the conifold locus

In order to apply the gap condition in this example, we have to transform and solve the Picard-Fuchs system at a specific point on the conifold locus. We make the choice $z_1 = 2$, $z_2 = -\frac{1}{2}$. Again we define two variables which vanish at this point

$$z_{c,1} = 1 - \frac{z_2}{-\frac{1}{4}(z_1-2)-\frac{1}{2}}, \quad z_{c,2} = 1 - \frac{z_2}{4(z_1-2)-\frac{1}{2}}. \qquad (5.4.18)$$

$z_{c,1}$ is a coordinate normal to the conifold divisor and $z_{c,2}$ describes a tangential direction. Transforming the Picard-Fuchs system to these coordinates we find the following set of periods:

$$\begin{aligned}
\omega_0^c &= 1, \\
\omega_1^c &= z_{c,1} + \frac{6773 z_{c,1}^2}{14450} - \frac{58 z_{c,1} z_{c,2}}{7225} - \frac{z_{c,2}^2}{1445} + \mathcal{O}(z_c^3), \\
\omega_2^c &= z_{c,2} + \frac{10858 z_{c,1}^2}{7225} + \frac{2871 z_{c,2}^2}{2890} - \frac{4886 z_{c,1} z_{c,2}}{7225} + \mathcal{O}(z_c^3).
\end{aligned} \qquad (5.4.19)$$

Next, we can express the $z_{c,i}$ through the mirror coordinates $t_{c,1} := \omega_1^c$ and $t_{c,2} := \omega_2^c$ by inverting the above series

$$\begin{aligned}
z_{c,1}(t_{c,1}, t_{c,2}) &= t_{c,1} - \frac{6773 t_{c,1}^2}{14450} + \frac{58 t_{c,1} t_{c,2}}{7225} + \frac{t_{c,2}^2}{1445} + \mathcal{O}(t_c^3), \\
z_{c,2}(t_{c,1}, t_{c,2}) &= t_{c,2} - \frac{10858 t_{c,1}^2}{7225} + \frac{4886 t_{c,1} t_{c,2}}{7225} - \frac{2871 t_{c,2}^2}{2890} + \mathcal{O}(t_c^3).
\end{aligned} \qquad (5.4.20)$$

Transforming the Yukawa couplings, the Christoffel symbols and the holomorphic ambiguities $\tilde{f}$ to the conifold coordinates we obtain the propagator around this locus. In the

5.4. $\mathbb{K}_{\mathbb{F}_1} = \mathcal{O}(-2,-3) \to \mathbb{F}_1$

choice of our coordinates the propagator takes the following form

$$S^{11} = \frac{5}{12} - \frac{2t_{c,1}}{25} - \frac{337t_{c,1}^2}{10625} - \frac{4t_{c,1}t_{c,2}}{2125} + \mathcal{O}(t_c^3),$$

$$S^{12} = -\frac{55}{4} + \frac{66t_{c,1}}{25} + \frac{11121t_{c,1}^2}{10625} + \frac{132t_{c,1}t_{c,2}}{2125} + \mathcal{O}(t_c^3), \qquad (5.4.21)$$

$$S^{22} = \frac{1815}{4} - \frac{2178t_{c,1}}{25} - \frac{366993t_{c,1}^2}{10625} - \frac{4356t_{c,1}t_{c,2}}{2125} + \mathcal{O}(t_c^3).$$

Again the gap condition in combination with the known leading behavior at the large radius point suffices to fix all coefficients in the holomorphic ambiguity. From the conifold alone we get at genus two 27 out of 28 unknowns and at genus three 90 out of 91 unknowns. Our results read

$$\begin{aligned} F_c^2 &= \frac{1}{48t_{c,1}^2} + \frac{1567}{9000000} + \frac{98333}{1593750000}t_{c,1} - \frac{123}{10625000}t_{c,2} + \mathcal{O}(t_c^2) \\ F_c^3 &= \frac{25}{1008t_{c,1}^4} + \frac{480217}{283500000000} + \frac{106245283t_{c,1}}{17929687500000} + \frac{69949t_{c,2}}{167343750000} + \mathcal{O}(t_c^2). \end{aligned} \qquad (5.4.22)$$

5.4.3 Relation to the family of elliptic curves

Starting point is again the j-function of $\tilde{\Sigma}(z_1, z_2)$ which we will repeat here

$$j(z_1, z_2) = \frac{((1 - 4z_1)^2 + 24z_1 z_2)^3}{z_1^3 z_2^2 ((1 - 4z_1)^2 - z_2(1 - 36z_1 + 27z_1 z_2))}. \qquad (5.4.23)$$

Using again the usual j-function description (B.1.11) one can establish a relation between the elliptic parameter $q = e^{2\pi i \tau}$ and the complex structure variables z_1 and z_2 which reads

$$q = z_1^3 z_2^2 + 16 z_1^4 z_2^2 + 160 z_1^5 z_2^2 - z_1^3 z_2^3 - 60 z_1^4 z_2^3 + \mathcal{O}(z^8). \qquad (5.4.24)$$

We observe that

$$\tau = \partial_{t_{x,2}} \partial_{t_{x,2}} F^0, \quad \partial_{t_{x,2}} \tau = -C_{t_{x,2}t_{x,2}t_{x,2}}, \qquad (5.4.25)$$

where $t_{x,i}$ is obtained from $Q_i^x = e^{2\pi i t_{x,i}}$, which hints at that the not instanton corrected parameter x_1 or Q_1^x, respectively, is merely an auxiliary parameter. As in the previous cases it is possible to write the full non-holomorphic $\mathcal{F}^{(1)}$ as

$$\mathcal{F}^{(1)} = -\log \sqrt{\tau_2} \eta(\tau) \bar{\eta}(\bar{\tau}) + A, \qquad (5.4.26)$$

where A is given by (5.4.16).

Chapter 6

K3 fibrations

In this chapter we construct solutions to the holomorphic anomaly equations for regular K3-fibrations with two moduli, which are realized as hypersurfaces in toric ambient spaces. From the perspective of the 2d σ model these geometries correspond to σ model with superpotential. We focus in particular on two models whose moduli spaces have been explored in detail [82]. These moduli spaces have identical types of boundary divisors, which have been interpreted physically in the context of heterotic/type II duality. Apart from the large radius point one has the weak coupling divisor, the strong coupling divisor, a Seiberg-Witten divisor, the generic conifold locus and the Gepner point where a conformal field theory description becomes available.

The material presented here was published in reference [5].

6.1 Calabi-Yau hypersurfaces in toric varieties

Compact toric ambient spaces $\mathbb{P}_\Sigma$ of complex dimension d are described through the quotient

$$\mathbb{P}_\Sigma = (\mathbb{C}^n - Z)/G, \qquad (6.1.1)$$

where $G \cong (\mathbb{C}^*)^h$ with $h = n - d$ and one has to exclude an exceptional set $Z \subset \mathbb{C}^n$ to obtain a well-behaved quotient. The h independent $\mathbb{C}^*$ identifications arise as follows. $\mathbb{P}_\Sigma$ is defined in terms of a fan Σ, which is a collection of rational polyhedral cones $\sigma \in \Sigma$ containing all faces and intersections of its elements [73, 74]. The cones are spanned by vectors which are sitting in a d-dimensional integral lattice Γ^* and $\mathbb{P}_\Sigma$ is compact if the support of Σ covers all of the real extension $\Gamma^*_\mathbb{R} = \Gamma^* \otimes \mathbb{R}$ of the lattice Γ^*. We will concentrate on the case where Σ consists of the cones over the faces of an integral polyhedron $\Delta^* \subset \Gamma^*_\mathbb{R}$, which contains the origin $v_0 = (0, \ldots, 0)$. In toric geometry l-dimensional cones of Δ^* represent codimension l subvarieties of $\mathbb{P}_\Sigma$. Now, let $\Sigma(1)$ denote the set of one-dimensional cones with primitive generators $v_i, i = 1, \cdots, n$. One finds that there are h n-vectors $Q_i^a \in \mathbb{Z}, a = 1, \cdots, h$, called charge vectors, satisfying the linear relations $\sum_{i=1}^n Q_i^a v_i = 0$ among the primitive lattice vectors v_i. This defines an action of

the group G on the homogeneous coordinates $x_i \in \mathbb{C}$ as follows: $x_i \mapsto \mu_a^{Q_i^a} x_i$ with $\mu_a \in \mathbb{C}^*$.
Anti-canonical hypersurfaces in $\mathbb{P}_\Sigma$ are given by sections of the anti-canonical bundle $\mathcal{O}_{\mathbb{P}_\Sigma}(\sum_{v_i \in \Sigma(1)} D_i)$, where D_i is the corresponding divisor to $v_i \in \Sigma(1)$. In order for these to be Calabi-Yau a further condition must be satisfied. $\mathbb{P}_\Sigma$ will usually have singularities which have to be blown up and the criterion for the canonical bundle to extend to a bundle of the blow-up is that the singularities are of Gorenstein type. Once we also require $\mathbb{P}_\Sigma$ to be Fano, i.e. that the anti-canonical bundle is positive, the above hypersurfaces will define Calabi-Yau hypersurfaces $M \subset \mathbb{P}_\Sigma$.

Let us now pass over to the description of mirror symmetry for these Calabi-Yau hypersurfaces. We denote by Γ the dual lattice to Γ^*, $\Gamma_\mathbb{R}$ the real extension and by $\langle \cdot, \cdot \rangle$ the canonical pairing between the dual vector spaces and define the dual polyhedron Δ as

$$\Delta = \{ m \in \Gamma_\mathbb{R} | \langle n, m \rangle \geq -1 \text{ for all } n \in \Delta^* \}. \tag{6.1.2}$$

If all vertices of Δ belong to Γ and Δ contains the origin, then Δ is again an integral polyhedron and both Δ^* as well as Δ are called reflexive. Note that this implies that in both Δ^* and Δ the origin is the only interior point. In [83] Batyrev showed that Δ is reflexive if and only if the corresponding toric variety, denoted by $\mathbb{P}_{\Sigma^*}$, is Gorenstein and Fano. This opens up the way for the construction of the mirror Calabi-Yau manifold as a hypersurface in $\mathbb{P}_{\Sigma^*}$, where Σ^* is the fan over the faces of Δ. The construction uses the fact that the toric variety corresponding to a fan Σ can be defined alternatively through the polyhedron Δ as an embedding $\mathbb{P}_\Sigma = \mathbb{P}_{\Sigma(\Delta)} \hookrightarrow \mathbb{P}^k$ with $k = |\Delta \cap \Gamma| - 1$ using the linear relations among the vertices of Δ. The same applies to the toric variety corresponding to Σ^*, where now $\mathbb{P}_{\Sigma^*} = \mathbb{P}_{\Sigma(\Delta^*)} \hookrightarrow \mathbb{P}^{k'}$ with $k' = |\Delta^* \cap \Gamma^*| - 1$. Batyrev showed that the mirror Calabi-Yau manifold W is given by the anti-canonical hypersurface in $\mathbb{P}_{\Sigma^*}$. The Hodge numbers can be computed through methods of toric geometry and one obtains [83]:

$$h^{1,1} = l(\Delta^*) - d - 1 - \sum_{\gamma^*} l^*(\gamma^*) + \sum_{\Theta^*} l^*(\Theta^*) l^*(\hat{\Theta}^*) \tag{6.1.3}$$

$$h^{d-2,1} = l(\Delta) - d - 1 - \sum_\gamma l^*(\gamma) + \sum_\Theta l^*(\Theta) l^*(\hat{\Theta}) . \tag{6.1.4}$$

Here γ^* (γ) refers to codimension 1 faces of Δ^* (Δ) and Θ^* (Θ) refers to codimension 2 faces of Δ^* (Δ). By $\hat{\Theta}^*$ we denote the face of Δ, which is dual to Θ^* in Δ^* and vice versa. If F is a facet of the polytop Δ or Δ^* then $l(F)$ denotes the set of all integral points on F, while $l^*(F)$ denotes only the interior integral points, i.e. those which do not lie in codimension one facets of F.

In the following we will describe the case $h = 1$ and $d = 4$, i.e. 3-dimensional hypersurfaces in weighted projective space, as this is the relevant construction for our particular models. We also assume that one weight is 1 and all weights divide the degree D of the anticanonical hypersurface and denote the weight vector by $(Q_1, Q_2, Q_3, Q_4, 1)$. Then Δ^* spanned by the vertices

$$v_1 = (1, 0, 0, 0), \quad v_2 = (0, 1, 0, 0), \quad v_3 = (0, 0, 1, 0), \quad v_4 = (0, 0, 0, 1),$$
$$v_5 = (-Q_1, -Q_2, -Q_3, -Q_4)$$

is an reflexive polyhedron and the charge vector is identified with the weight vector. It is convenient to consider also the extended vertices $\bar{v}_i = (1, v_i)$. The linear relation between the extended vertices $\sum_{i=1}^{5} Q_i \bar{v}_i = D \bar{v}_0$ reproduces the Calabi-Yau condition for $c_1(\mathcal{T}_M) = 0$ namely $D = \sum_i Q_i$, where d is the degree of the hypersurface.

According to the above description this on the one hand defines the toric variety $\mathbb{P}_\Sigma$ as the weighted projective space $\mathbb{P}_d^{(\vec{Q})}$ with the family of Calabi-Yau hypersurfaces M given by generic degree D homogeneous polynomials and we write $M = \mathbb{P}_d^{(\vec{Q})}[D] \subset \mathbb{P}_\Sigma$. On the other hand the linear dependence of the charge vectors leads to the identification

$$\mathbb{P}_{\Sigma^*} = \mathbb{P}_{\Sigma(\Delta^*)} \equiv \mathbf{H}^5(\vec{Q}) := \{(U_0, U_1, U_2, U_3, U_4, U_5) \in \mathbb{P}_5 | \prod_{i=1}^{5} U_i^{Q_i} = U_0^D\}. \quad (6.1.5)$$

One can now consider the embedding map $\phi : \mathbb{P}_4^{(\vec{Q})} \to \mathbf{H}^5(\vec{Q})$ given by

$$[y_1, y_2, y_3, y_4, y_5] \mapsto [y_1 y_2 y_3 y_4 y_5, y_1^{D/Q_1}, y_2^{D/Q_2}, y_3^{D/Q_3}, y_4^{D/Q_4}, y_5^{D/Q_5}], \quad (6.1.6)$$

which defines the isomorphism $\mathbb{P}_{\Sigma^*} \cong \mathbb{P}_4^{(\vec{Q})}/\mathrm{Ker}\,\phi$. Anti-canonical hypersurfaces in $\mathbb{P}_{\Sigma^*}$ are defined through expressions linear in the U_i which in turn can be expressed as monomials in the y_i through equation (6.1.6). Resolution of the singularities arising from $\mathbb{P}_{\Sigma^*}$ then gives the family of mirror Calabi-Yau hypersurfaces $W \subset \mathbb{P}_{\Sigma^*}$.

6.2 Picard-Fuchs equations and the B-model

We want to analyze the periods of the mirror Calabi-Yau. The mirror is given by sections of the anti-canonical bundle of $\mathbb{P}_{\Sigma^*}$. These can be identified with the Laurent polynomials

$$f = \sum_i a_i Y^{m_i}, \quad Y^{m_i} = Y_1^{m_i^1} Y_2^{m_i^2} \cdots Y_d^{m_i^d}, \quad (6.2.1)$$

where $(Y_1, \cdots, Y_d)$ are coordinates for the torus $T \subset \mathbb{P}_{\Sigma^*}$ and m_i are points of $\Delta^* \cap \Gamma^*$ which do not lie in the interior of codimension one faces of Δ^*. The Griffiths construction [84] then gives the following set of Periods:

$$\Pi_i(a) = \int_{\gamma_i} \Omega = \int_{\gamma_i} \frac{1}{f(a, Y)} \prod_{j=1}^{d} \frac{dY_j}{Y_j}, \quad (6.2.2)$$

with $\gamma_i \in H_d((\mathbb{C}^*)^d \backslash Z_f)$. Here Z_f is the vanishing locus of the polynomial f. These periods satisfy a set of differential equations which are called the GKZ system [85]. In order to obtain them, we extend the vectors v_i from above to the $\mathbb{R}^{d+1}$ dimensional vectors $\bar{v}_i = (1, v_i)$ forming the set $A = \{\bar{v}_0, \cdots, \bar{v}_n\}$. Assuming that these integral points span $\mathbb{Z}^{d+1}$ we obtain $h = n - d$ linear dependencies described by the lattice

$$\Lambda = \{(l_0^{(k)}, \cdots, l_n^{(k)}) \in \mathbb{Z}^{n+1} | \sum_{i=0}^{n} l_i^{(k)} \bar{v}_i = 0, \ k = 1, \ldots, h\}. \quad (6.2.3)$$

Now we are ready to write down the differential operators which annihilate the periods (6.2.2):

$$\mathcal{D}_k = \prod_{l_i^{(k)}>0} \left(\frac{\partial}{\partial a_i}\right)^{l_i^{(k)}} - \prod_{l_i^{(k)}<0} \left(\frac{\partial}{\partial a_i}\right)^{-l_i^{(k)}}, \qquad (6.2.4)$$

for each element $l^{(k)}$ of Λ and

$$\mathcal{Z}_j = \sum_{i=0}^{n} \bar{v}_{i,j} a_i \frac{\partial}{\partial a_i} - \beta_j, \qquad (6.2.5)$$

where $\beta \in \mathbb{R}^{d+1}$ and $\bar{v}_{i,j}$ represent the j-th component of the vector $\bar{v}_i \in \mathbb{R}^{d+1}$. One can show that equation (6.2.5) defines invariance under the rescalings $a_i \mapsto \lambda_j^{m_{i,j}} a_i$ and $f \mapsto c \cdot f$ for $\lambda, c \in \mathbb{C}^*$. Therefore we define the invariant variable

$$z_j = (-1)^{l_0^{(k)}} \prod_i a_i^{l_i^{(j)}}, \qquad (6.2.6)$$

which transforms (6.2.4) to a generalized system of hypergeometric equations

$$\mathcal{D}_k \Pi_i(z_1, \cdots, z_{|\Lambda|}) = 0 \qquad (6.2.7)$$

for each $l^{(k)} \in \Lambda$.

In general, these differential equations forming the so called $\mathcal{A}$-system, contain the periods among their solutions, but there will be also other solutions. The set of Picard-Fuchs equations which vanish only on the periods can be obtained by factoring the above equations. Then the resulting lower order operator is Picard-Fuchs once it annihilates all periods.

For a general set of Picard-Fuchs equations, the solution space has dimension $h_3(W)$ and one obtains the following set of periods [86]:

$$\Pi(z) = \begin{pmatrix} \omega_0(z,\rho)|_{\rho=0} \\ D_i^{(1)} \omega_0(z,\rho)|_{\rho=0} \\ D_i^{(2)} \omega_0(z,\rho)|_{\rho=0} \\ D^{(3)} \omega_0(z,\rho)|_{\rho=0} \end{pmatrix}. \qquad (6.2.8)$$

Here, i runs from 1 to $h_{21}(W)$, where $h_{21}(W)$ is the number of moduli. Furthermore we have the following definitions:

$$\omega_0(z,\rho) = \sum_{n_i \geq 0} c(n+\rho) z^{n+\rho}, \qquad (6.2.9)$$

$$D_i^{(1)} := \partial_{\rho_i}, \ D_i^{(2)} := \frac{1}{2} \kappa_{ijk} \partial_{\rho_j} \partial_{\rho_k}, \ D^{(3)} := -\frac{1}{6} \kappa_{ijk} \partial_{\rho_i} \partial_{\rho_j} \partial_{\rho_k}, \qquad (6.2.10)$$

where κ_{ijk} are the classical intersection numbers of the Calabi-Yau M and $c(n+\rho)$ is defined by

$$c(n+\rho) = \frac{\Gamma\left(\sum_{k=1}^{h} l_0^{(k)}(n_k + \rho_k) + 1\right)}{\prod_{i=1}^{n} \Gamma\left(\sum_{k=1}^{h} l_i^{(k)}(n_k + \rho_k) + 1\right)} . \tag{6.2.11}$$

As described in section 3.6 the periods (6.2.8) describe complex structure deformations of the manifold W and can be written in terms of homogeneous special coordinates (X^0, X^i) as $(X^0, X^i, (\partial \mathcal{F}/\partial X^i), (\partial \mathcal{F}/\partial X^0))$. With the choice (6.2.10) they are already in the integer symplectic basis and thus correspond directly to the A model period vector $\Pi(t) = (1, t^i, \partial_i F, 2F - t^i \partial_i F)$ multiplied with X^0.

Then the mirror map takes the form

$$t^i(z) = \frac{\omega_i(z)}{\omega_0(z)}, \quad \omega_i(z) := D_i^{(1)} \omega_0(z, \rho)|_{\rho=0} . \tag{6.2.12}$$

6.3 Moduli Space of K3 Fibrations

In this section we want to give an overview of the moduli Space of K3 fibrations as presented in [82]. Special attendance will be given to boundary divisors and their importance for physical boundary conditions.

6.3.1 K3 Fibrations

K3 fibrations arise in the context of heterotic/type II duality once one wants to have $N = 2$ supersymmetry in four dimensions [87]. In order to achieve this amount of supersymmetry on the heterotic side one has to compactify on $K3 \times T^2$. In the heterotic picture vector multiplets come from the 2-torus together with its bundle and the dilaton axion, while hypermultiplets arise as deformations of the K3 surface and its bundle. On the type II side vector multiplets arise from compactification of the R-R three-form and therefore count $h^{1,1}(M)$, while the complex structure deformation parameters of the Calabi-Yau together with the dilaton-axion form $h^{2,1}(M) + 1$ hypermultiplets. Part of the duality conjecture is that there is a complete match between the moduli spaces of the two theories. To see the consequences it is easiest to start with the vector multiplet moduli of the heterotic side. Here one has from the Narain moduli of the 2-torus and the dilaton-axion locally a product of the form

$$\frac{O(2, m)}{O(2) \times O(m)} \times \frac{SL(2)}{U(1)} . \tag{6.3.1}$$

The classical vector moduli space $\mathcal{M}_V$ of the type IIA theory is a special Kähler manifold. Choosing special coordinates, the Kähler potential takes the following form in terms of the prepotential F

$$K = -\log\left(2(F+\bar{F}) - (t^i - \bar{t}^i)\left(\frac{\partial F}{\partial t^i} - \frac{\partial \bar{F}}{\partial \bar{t}^i}\right)\right),$$

$$G_{i\bar{j}} = \frac{\partial K}{\partial t^i \partial \bar{t}^j}. \qquad (6.3.2)$$

Here, $G_{i\bar{j}}$ is the metric on moduli space. Imposing the local product structure (6.3.1) on a special Kähler manifold allows one to deduce the prepotential from equations (6.3.2). This, on the other hand, opens up the way for calculating intersection numbers $\kappa_{ijk} = \sharp(D_i \cap D_j \cap D_k)$ from the special Kähler relation

$$\kappa_{ijk} = \frac{\partial F}{\partial t_i \partial t_j \partial t_k}. \qquad (6.3.3)$$

Such a calculation was performed in [88] and the result is

$$\begin{aligned}\sharp(D_0 \cap D_0 \cap D_0) &= 0, \\ \sharp(D_0 \cap D_0 \cap D_i) &= 0, \quad i = 1, \cdots, h^{1,1} - 1 \\ \sharp(D_0 \cap D_i \cap D_j) &= \eta_{ij}, \quad i,j = 1, \cdots, h^{1,1} - 1,\end{aligned} \qquad (6.3.4)$$

where η_{ij} is a matrix of nonzero determinant and signature $(+, -, -, \cdots, -)$. A theorem of Oguiso [89] says that if M is a Calabi-Yau threefold and L a divisor such that

$$L \cdot c \geq 0 \text{ for all curves } c \in H_2(M, \mathbb{Z}), \quad L^2 \cdot D = 0 \text{ for all divisors } D \in H_4(M, \mathbb{Z}), \quad (6.3.5)$$

then there is a fibration $\Phi : M \to W$, where W is $\mathbb{P}^1$ and the generic fibre L is either a K3 surface or an abelian surface. The second condition of (6.3.5) follows from (6.3.4) with $L = D_0$. Further investigation [90] shows that the first condition in (6.3.5) is also true and that the Euler characteristic of the fibre given by the second chern class is 24. This determines the fibration as a K3 fibration over $\mathbb{P}^1$, where the first factor of (6.3.1) arises from the Picard-lattice of the K3 fibre and the second factor is identified with the Kähler form plus B-field on $\mathbb{P}^1$. Therefore we see that in this picture the heterotic dilaton is identified with the size of the $\mathbb{P}^1$ on the type IIA side.

6.3.2 The Moduli Space of the Mirror

In this section we will concentrate on the example $M_1 = \mathbb{P}_4^{(1,1,2,2,2)}[8]$ whose mirror moduli space we will describe. The case of $M_2 = \mathbb{P}_4^{(1,1,2,2,0)}[12]$ is analogous. M_1 is a generic degree 8 hypersurface in the weighted projective space $\mathbb{P}_4^{(1,1,2,2,2)}$. A typical defining polynomial for such a hypersurface is

$$p = x_1^8 + x_2^8 + x_3^4 + x_4^4 + x_5^4. \qquad (6.3.6)$$

One sees that there is a $\mathbb{Z}_2$-singularity along $[0, 0, x_3, x_4, x_5]$. This may be blown up, replacing each point in the locus by $\mathbb{P}^1$ with homogeneous coordinates $[x_1, x_2]$ which will

6.3. MODULI SPACE OF K3 FIBRATIONS

be the base of the fibration. Choosing a point on the base by fixing $x_1/x_2 = \lambda$ projects onto the subspace $\mathbb{P}_3^{(1,2,2,2)}$ with the fibre given by the hypersurface

$$(\lambda^8 + 1)x_2^8 + x_3^4 + x_4^4 + x_5^4 = 0. \tag{6.3.7}$$

This is seen to be a quartic K3 surface in $\mathbb{P}_3$ ones one makes the substitution $y_1 = x_2^2$. Thus we see that M_1 is a K3 fibration with two Kähler moduli t_1 and t_2, t_2 being the size of the $\mathbb{P}^1$ base while t_1 corresponds to the curve cut out by a generic hyperplane in the K3 fibre. Its mirror W_1 may be identified with the family of Calabi-Yau threefolds of the form $\{p = 0\}/G$, where

$$p = x_1^8 + x_2^8 + x_3^4 + x_4^4 + x_5^4 - 8\psi x_1 x_2 x_3 x_4 x_5 - 2\phi x_1^4 x_2^4. \tag{6.3.8}$$

Here, G consists of elements $g = (\alpha^{a_1}, \alpha^{a_2}, \alpha^{2a_3}, \alpha^{2a_4}, \alpha^{2a_5})$ with the action

$$(x_1, x_2, x_3, x_4, x_5; \psi, \phi) \mapsto (\alpha^{a_1} x_1, \alpha^{a_2} x_2, \alpha^{2a_3} x_3, \alpha^{2a_4} x_4, \alpha^{2a_5} x_5; \alpha^{-a} \psi, \alpha^{-4a} \phi), \tag{6.3.9}$$

where $a = a_1 + a_2 + 2a_3 + 2a_4 + 2a_5$, where α^{a_1} and α^{a_2} are 8th roots of unity, and where α^{2a_3}, α^{2a_4}, and α^{2a_5} are 4th roots of unity. Therefore, we see that the parameter space $\{(\psi, \phi)\}$ is modded out by a $\mathbb{Z}_8$ acting in the form

$$(\psi, \phi) \mapsto (\alpha \psi, -\phi). \tag{6.3.10}$$

This translates to the description of the moduli space as an affine quadric in $\mathbb{C}^3$

$$\tilde{\xi}\tilde{\zeta} = \tilde{\eta}^2, \tag{6.3.11}$$

with invariant coordinates

$$\tilde{\xi} = \psi^8, \quad \tilde{\eta} = \psi^4 \phi, \quad \tilde{\zeta} = \phi^2. \tag{6.3.12}$$

Compactification of this space leads to the projective quadric $Q = \{\xi\zeta - \eta^2 = 0\}$ in $\mathbb{P}_3$ with coordinates $[\xi, \eta, \zeta, \tau]$. The relation to the tilted coordinates is $\tilde{\xi} = \xi/\tau$, $\tilde{\eta} = \eta/\tau$, $\tilde{\zeta} = \zeta/\tau$ for $\tau \neq 0$. Having identified the global form of the moduli space let us now proceed to the description of boundary divisors which correspond to parameter values for which the original family of hypersurfaces $\{p = 0\}$ develops singularities. The analysis was done in [82] and the result is the following set of boundary divisors:

$$C_{con} = Q \cap \{1 - 2x + x^2(1-y) = 0\}, \tag{6.3.13}$$
$$C_1 = Q \cap \{1 - y = 0\}, \tag{6.3.14}$$
$$C_\infty = Q \cap \{y = 0\} : \quad \phi \text{ and } \psi \text{ both approach infinity}, \tag{6.3.15}$$
$$C_0 = Q \cap \{1/x = 0\}. \tag{6.3.16}$$

Here we have used the coordinates [1]

$$x := -\frac{1}{8}\phi\psi^{-4}, \quad y = \phi^{-2}, \tag{6.3.17}$$

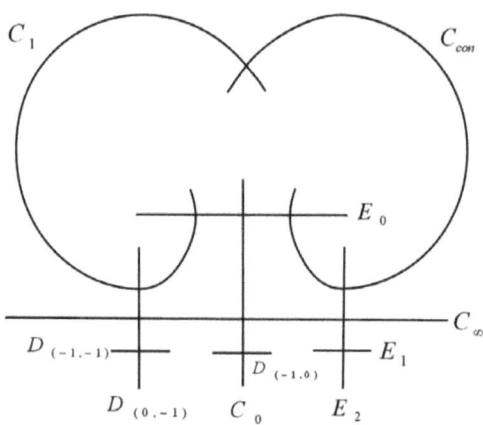

Figure 6.1: The blown up moduli space.

which are themselves related to the coordinates z_i in (6.2.6) through $z_1 = x/256$ and $z_2 = y/4$. This choice is convenient for the description of the mirror map which becomes

$$t_1 \sim \log(x), \quad t_2 \sim \log(y). \tag{6.3.18}$$

C_{con} corresponds to the locus where the Calabi-Yau developers a conifold singularity and along C_1 the Calabi-Yau manifold M_1 admits a whole singular curve of genus g over which A_{n-1} singularities are fibred (in our particular case $g = 3$ and $n = 2$). C_∞ is the locus where the volume of the $\mathbb{P}^1$ base goes to infinity. Next, one notices that the resulting space is singular. First of all the quadric $Q \subset \mathbb{P}^3$ is singular by itself. This singularity is of toric origin as Q can be identified isomorphically with $\mathbb{P}^{1,1,2}$. Further singularities arise from the point of tangency between the divisors C_{con}, C_∞, from the tangency between C_1, C_∞ (of toric origin), and from the common point of intersection of C_0, C_1 and C_{con}. Blowing up all singular points leads to the schematic picture of the moduli space presented in figure 6.1.

6.4 Physical boundary conditions

6.4.1 The Strong coupling singularity

Consider the locus $C_1 = Q \cap \{\phi^2 = 1\}$. It can be shown that the mirror map converts this locus to the locus $t_2 = 0$ in the Kähler moduli space of the Calabi-Yau M_1 [91]. As t_2

---

[1] For the model $\mathbb{P}_4^{(1,1,2,2,6)}$ [12] use $x := -\frac{1}{864}\frac{\phi}{\psi^6}$, $y = \frac{1}{\phi^2}$.

6.4. PHYSICAL BOUNDARY CONDITIONS

describes the size of the $\mathbb{P}^1$ which is the base of the K3 fibration $t_2 = 0$ translates to the strong coupling regime in the dual heterotic picture. In M_1 the singularity is described by the equations

$$x_1 = x_2 = 0, \quad x_3^4 + x_4^4 + x_5^4 = 0, \tag{6.4.1}$$

leading to a genus 3 curve [2] C of fixed points of the projective action $x_i \mapsto \mu^{Q_i} x_i$. In the language of toric geometry the singular curve C corresponds to a one-dimensional edge of the dual polyhedron Δ^* with integral lattice points on it. The resolution process adds a new vertex for each of these points leading to an exceptional $\mathbb{P}^1$ bundle over C in the blown up of the Calabi-Yau manifold for each ray added. The monomial divisor mirror map relates each vertex to the addition of a new perturbation in the defining polynomial of W_1. In our case we blow up only once and the perturbation added is the term $\phi x_1^4 x_2^4$ in (6.3.8).

To see what happens from the physics point of view along C_1 we look at the effective action arising from compactification in the type IIA picture. This procedure has been analyzed in [92]. Let us first clarify the setup. Assume that we have a smooth curve of genus g and singularities of type A_{N-1} fibred over the curve. The resolution of the transverse A_{N-1} singularity gives rise to an ALE space in which the vanishing cycles are described by a chain of $N-1$ two-spheres Γ_i and their intersection matrix corresponds to the Dynkin diagram of A_{N-1}. Now consider soliton states described by two-branes wrapping the two-cycles A^{ij} defined by the chain $\Gamma_i \cup \Gamma_{i+1} \cup \cdots \cup \Gamma_j$. These become charged under $U(1)^{N-1}$ with their charges being identified with the positive roots of A_{N-1}. Compactification of the theory down to 4 dimensions leads to a $N=2$ supersymmetric $SU(N)$ gauge theory with g hypermultiplets (coming from holomorphic 1-forms on C) transforming in the adjoint representation of the gauge group. In $N=1$ superfield notation, we obtain the following effective Lagrangian

$$2\pi \mathcal{L} = \operatorname{Im}\left[\operatorname{Tr} \int d^4\theta (M_i^\dagger e^V M^i + \tilde{M}^{i\dagger} e^V \tilde{M}_i + \Phi^\dagger e^V \Phi) + \frac{\tau}{2}\int d^2\theta \operatorname{Tr} W^2 + i \int d^2\theta \mathcal{W}\right], \tag{6.4.2}$$

with the superpotential

$$\mathcal{W} = \operatorname{Tr}\tilde{M}^i[\Phi, M_i], \tag{6.4.3}$$

and the scalar potential

$$\begin{aligned}\mathcal{V} &= \operatorname{Tr}\left[[m_i, m^{\dagger i}]^2 + [\tilde{m}^i, \tilde{m}_i^\dagger]^2 + [\phi, \phi^\dagger]^2 \right.\\ &\left. + 2\left([m^{\dagger i}, \phi][\phi^\dagger, m_i] + [\tilde{m}_i^\dagger, \phi][\phi^\dagger, \tilde{m}^i] + [m_i, \tilde{m}_i^\dagger][\tilde{m}_j^\dagger, m^{\dagger j}]\right)\right].\end{aligned} \tag{6.4.4}$$

Here V_a is the vector multiplet in the adjoint and W_a^a its field strength. Furthermore, one has a chiral superfield Φ^a in the adjoint (comprising with V the $N=2$ vector multiplet), and $2g$ chiral superfields M_a^i, $\tilde{M}_a^i$ in the adjoint (comprising the g hypermultiplets, $i = 1, \cdots, g$). Going to the Coulomb branch of the moduli space, $\phi = \operatorname{diag}(\phi_1, \phi_2, \cdots, \phi_N)$

---

[2] In M_2 the equations describing the singularity lead to a genus 2 curve

with $\sum \phi_i = 0$, one sees that at generic points along this the gauge symmetry is spontaneously broken to $U(1)^{N-1}$. In codimension one (along the singular divisor $C_1 \sim \phi_N = 0$) the unbroken symmetry gets enhanced to $SU(2) \times U(1)^{N-2}$ and from (6.4.4) one can deduce that $2g$ hypermultiplets and 2 vector multiplets are becoming massless near the locus $\phi_N = 0$. Therefore, the number $n_H - n_V$ in (3.7.28) becomes $2g - 2$.

6.4.2 The weak coupling divisor and meromorphic modular forms

The weak coupling divisor deserves its name from the definition $y = 0$ which inserted into the mirror map (6.3.18) gives $t_2 \to \infty$. Again as t_2 describes the size of the dilaton in the heterotic dual we are in the weak coupling regime of the heterotic string. This can be used to calculate higher genus amplitudes in the type IIA string through a heterotic one loop computation. As was explained in section 3.4.3 the topological free energies F^g compute the moduli dependent coupling of graviton-graviphoton scatterings. F^g is a homogeneous function of the homogeneous coordinates X^I of degree $2 - 2g$ and X^0 can be chosen to be

$$X^0 = \frac{1}{g_s} e^{K/2}, \qquad (6.4.5)$$

one can write

$$F^g(X) = (X^0)^{2-2g} F^g(t) = (g_s^2)^{g-1} e^{(1-g)K} F^g(t). \qquad (6.4.6)$$

In type IIA string theory the Kähler potential K is independent of the dilaton since the latter belongs to a hypermultiplet and there are no neutral couplings between vector multiplets and hypermultiplets. The same argument tells us that $F^g(t)$ is independent of the dilaton and it follows from (6.4.6) that the couplings (3.4.25) appear only at genus g. Switching to the heterotic picture this statement changes as follows. Now the Kähler potential contains a $\log(g_s^2)$ term. This term arises from the vector moduli prepotential of the type IIA theory

$$F \sim ST^2 + \sum_{n=0}^{\infty} f_n(T) \exp(-nS) \qquad (6.4.7)$$

with the identifications $T = t_1$, $S = t_2$ and the choice $S = \frac{\theta}{2\pi} + i\frac{8\pi}{g_s^2}$. Next, notice that this implies that X^0 is of order 1 in the dilaton and therefore one extracts from (6.4.6) that all F^g appear at one loop in the heterotic theory. However, in the case of the heterotic string the dilaton belongs to a vector multiplet and therefore all $F^g(t)$ can have nontrivial dilaton dependence apart from the dependence through X^0. This implies that the above analysis is only valid in the limit $S \to \infty$ where the dilaton dependence in the $F^g(t)$ drops out. Translated to the type II picture we see that the one loop calculation gives only control over the terms in $F^g(t)$ which are independent of the class of the $\mathbb{P}^1$ base t_2. Such a calculation was performed in [94, 95] and extended to arbitrary regular K3 fibrations in [96]. The result is [96] that the Gopakumar-Vafa invariants for the K3 fibre are encoded in the following generating function

$$F_{K3}(\lambda, q) = \frac{2\Phi_{N,n}(q)}{q} \left(\frac{1}{2\sin(\frac{\lambda}{2})}\right)^2 \prod_{n \geq 1} \frac{1}{(1 - e^{i\lambda} q^n)^2 (1 - q^n)^{20} (1 - e^{-i\lambda} q^n)^2}, \qquad (6.4.8)$$

6.4. PHYSICAL BOUNDARY CONDITIONS

where $q = e^{2\pi i \tau}$ and $\Phi_{N,n}(q)$ is a modular form of half integral weight with respect to a congruent subgroup of $PSL(2,\mathbb{Z})$ acting in the standard form on τ. Let us review the essential points of this formula before we specify $\Phi_{N,n}(q)$ explicitly for the relevant class of K3 fibrations.

The formula applies to multi parameter K3 fibrations such as the 3 parameter STU model, as the Gopakumar-Vafa invariant depends on the class $[C]$ of the curve only via the self intersection C^2, and the latter is related to the exponents of the parameter q.

$\Phi_{N,n}$ is fully determined by the genus zero Gopakumar-Vafa invariants of the fibre direction. As it has been pointed out in [97], it can be also determined by classical geometric properties of the fibration, namely the embedding

$$\iota : \text{Pic}(K3) \hookrightarrow H^2(M,\mathbb{Z}), \qquad (6.4.9)$$

and the Noether-Lefshetz numbers of regular, i.e. non singular, one parameter families of quasi-polarized K3 surfaces $\pi : X \to \mathcal{C}$, where $\mathcal{C}$ is a curve. Let L be a quasi-polarization of degree

$$\int_{K3} L^2 = 2N. \qquad (6.4.10)$$

Then the family π yields a morphism $\imath : \mathcal{C} \to \mathcal{M}_{2N}$ to the moduli space of quasi-polarized K3 surfaces of degree $2N$. The Noether-Lefshetz numbers are defined by the intersection of $\mathcal{C}$ with the Noether-Lefshetz divisors in $\mathcal{M}_{2N}$. The latter are the closure of the loci in $\mathcal{M}_{2N}$ where the the rank of the Picard lattice is two. If β is an additional class in $\text{Pic}(K3)$ the Noether-Lefshetz divisor $D_{h,d} \in \mathcal{M}_{2N}$ may be labeled by $\int_{K3} \beta^2 = 2h-2$ and $\int_{K3} \beta \cdot L = d$ and combined into a generating function. The seminal work of Borcherds [98] relates these generating functions of the Noether-Lefshetz numbers to modular forms using the relations between Heegner- and Noether-Lefshetz divisors. In fact one knows that they are combinations of meromorphic vector valued modular forms of half integral weight. The theory of Borcherds can be viewed as a further extension of the work of Hirzebruch and Zagier on the modularity of counting functions of divisors in Hilbert modular surfaces.

In [96] a formula for $\Phi_{N,n}(q)$ of weight $\frac{21}{2}$ was found for regular K3 fibration Calabi-Yau, if the fibre is a quartic in $\mathbb{P}^3$, i.e. $N=2$, and if the fibre is a sixtic in the weighted projective space $\mathbb{P}(1,1,1,3)$, i.e. $N=1$. In fact there is a general formula, which incorporates not only the N parameter, but also a second parameter n parameterizing the different embedding (6.4.9). For the $N=1$ examples we have, [99],

$$\Phi_{1,n}(q) = U E_4 \left(\frac{U^4 \,(39\, V^8 + 26\, U^4\, V^4 - U^8)}{2^5} + n \frac{V^4\,(7\, U^8 - 6\, U^4\, V^4 - V^8)}{2^7} \right) \qquad (6.4.11)$$

with $U = \theta_3(\frac{\tau}{2})$, $V = \theta_4(\frac{\tau}{2})$ in terms of Jacobian theta functions

$$\theta_2 = \sum_{n=-\infty}^{\infty} q^{\frac{1}{2}(n+1/2)^2}, \qquad \theta_3 = \sum_{n=-\infty}^{\infty} q^{\frac{1}{2}n^2}, \qquad \theta_4 = \sum_{n=-\infty}^{\infty} (-1)^n q^{\frac{1}{2}n^2} \qquad (6.4.12)$$

and $E_4 = 1 - 240q + \ldots$ is the weight 4 Eisenstein series. According to (6.4.8) this has the following q expansion for the genus zero invariants

$$\frac{2\Phi_{1,n}}{\eta^{24}}(q^4) = \frac{2}{q^4} - 252 - 2496q - 223752q^4 - 725504q^5 - 15530000q^8 - 38637504q^9 \ldots$$
$$+ n\,(q^{-3} - 56 + 384q - 15024q^4 + 39933q^5 - 523584q^8 + 1129856q^9 \ldots)$$
(6.4.13)

and the coefficients of the $q^{d^2/N}$ are the genus zero BPS numbers n_d^0. We note in particular that the constant term is known from physical arguments and enumerative geometry to be the Euler number of the Calabi-Yau, i.e.

$$\chi = -252 - n\,56\,. \tag{6.4.14}$$

The fibrations discussed in this paper belong to the $n = 0$ case, but several manifolds with values $n \in \mathbb{Z}$ are realized as complete intersections or hypersurfaces in toric ambient spaces.

For the second type of fibrations that we treat in this paper with $N = 2$ one has

$$\Phi_{2,n} = \frac{1}{2^{21}}(81\,U^{19}V^2 - 3\,U^{21} + 627\,U^{18}V^3 + 14436\,U^{17}V^4 + 20007\,U^{16}V^5 + 169092\,U^{15}V^6$$
$$+ 120636\,U^{14}V^7 + 621558\,U^{13}V^8 + 292796\,U^{12}V^9 + 1038366\,U^{11}V^{10}$$
$$+ 346122\,U^{10}\,V^{11} + 878388\,U^9\,V^{12} + 207186\,U^8\,V^{13} + 361908\,U^7\,V^{14}$$
$$+ 56364\,U^6V^{15} + 60021\,U^5V^{16} + 4812\,U^4V^{17} + 1881\,U^3V^{18} + 27\,U^2V^{19} - V^{21})$$
$$- \frac{n}{2^{22}}U\,V\,(U^2 - V^2)^4(U^{11} - 21\,U^{10}\,V - 43\,U^9V^2 - 297\,U^8V^3 - 158\,U^7V^4 - 618\,U^6V^5$$
$$- 206\,U^5V^6 - 474\,U^4V^7 - 99\,U^3V^8 - 129\,U^2V^9 - 7\,U\,V^{10} + 3\,V^{11})$$
(6.4.15)

with $U = \theta_3(\frac{\tau}{4})$ and $V = \theta_4(\frac{\tau}{4})$ so that

$$\frac{2\Phi_{2,n}}{\eta^{24}}(q^8) = \frac{2}{q^8} - 168 - 640q - 10032q^4 - 158436q^8 - 288384q^9 - 1521216q^{12} - 10979984q^{16} + \ldots$$
$$+ n\left(\frac{2}{q^4} - 28 + 64q - 328q^4 - 1808q^8 + 2624q^9 - 7656q^{12} - 27928q^{16} \ldots\right)\,.$$
(6.4.16)

Again the genus zero invariants can be read of from the $q^{\frac{d^2}{N}}$ term and the $d = 0$ coefficient is related to the Euler number

$$\chi = -168 - n28\,. \tag{6.4.17}$$

One difficulty with the above approach is that due to η^{24} in the denominator one has to make an ansatz for the $\Gamma(4N)$ modular forms in the numerator of very high weight, as it is apparent in formula (6.4.16), while the quotient is always in $\mathcal{M}^!_{-\frac{3}{2}}(\Gamma_0(4N))$ and can be represented within a much smaller ring, as was pointed out by Zagier. The shriek stands for meromorphic, i.e. these forms are allowed to have arbitrary pole order at the cusp at $\tau \to i\infty$.

Such half integral, meromorphic forms are denoted by $\mathcal{M}^!_{k+\frac{1}{2}}(\Gamma_0(4N))$ with $k \in \mathbb{Z}$. They can be labeled by their weight and the pole order at the cusp. For a given weight there are forms of arbitrary pole orders, but our knowledge of the maximal pole order

6.4. PHYSICAL BOUNDARY CONDITIONS

keeps the problem finite. For $N = 1$ one can start with $\theta = f^{1,0}_{\frac{1}{2}} = \theta_3(2\tau)^3$ and for $N > 1$ one has to use the right combination of vector valued half integral forms which transform under the metaplectic representation of $\Gamma_0(4N)$. Let

$$\phi_k = \sum_{m=-\infty}^{\infty} q^{2Nm+k}, \qquad k = 0, \ldots, N \qquad (6.4.18)$$

then we define

$$\theta = \phi_0 + \phi_N + \frac{1}{2}\sum_{k=1}^{N-1} \phi_k . \qquad (6.4.19)$$

Different weights and pole orders can be constructed systematically by the following operations [100]:

- Multiplying f_r with $j(4N\tau)$, where j is the total modular invariant of $\Gamma_0 = PSL(2, \mathbb{Z})$, keeps the weight and shifts the pole order of the cusp at $i\infty$ by $-4N$.

- Taking derivatives $f_{r+2} = f_r E_2(4N\tau) - \frac{3}{Nr}f'_r(\tau)$ will shift the weight by 2 but keeps the pole order.

- Take the first Rankin-Cohen bracket $f_{k+l+2} = [f_k, E_l(4N)] = kf_k E'_l(4N) - lE_l(4N)f'_k$, where $E_4, E_6, E_8 = E_4^2$, etc. are holomorphic modular forms of Γ_1, shifts the weight by $l + 2$ but keeps the pole order. The bracket is build so that it cancels the inhomogeneous transformation of the derivatives

$$D = \frac{1}{2\pi i}\frac{\mathrm{d}}{\mathrm{d}\tau} . \qquad (6.4.20)$$

We denote $f' := Df$ etc. Similarly the nth rank Rankin-Cohen bracket of modular forms f, g of weight k, l is

$$[f, g]_n = \sum_{r=0}^{n}(-1)^r \binom{k+n-1}{n-r}\binom{l+n-1}{r} D^r f D^{n-r} g . \qquad (6.4.21)$$

and is modular of weight $k + l + 2n$. Choosing $f = f_r$ and g a holomorphic modular form changes the weight, but keeps the pole order.

- Dividing by $\Delta(4N\tau) = q^{4N}\prod_{m=1}^{\infty}(1-(q^{4N})^m)^{24}$ lowers the pole order by $-(4N)$ and the weight by -12.

It is clear that $\frac{2\Phi_{1,n}}{\eta^{24}}(q^4)$ must be of the form $f^{1,4}_{-\frac{3}{2}}(q^4)$, which can be build as follows. First we construct $u^{1,0}_{\frac{5}{2}} = \theta E_2(4\tau) - 6\theta'(\tau) = 1 - 10q - 70q^4 - 48q^5 + \ldots$. Form this

---

[3] Let us introduce the notation $f^{N,p}_r$, where we denote with r the weight, with p the pole order and N labels the congruence subgroup as above. We reserve the character f for forms, which are of the form $f^{N,p}_r = \frac{1}{q^p} + reg.$, while forms denoted by other characters can have subleading poles.

we can get two elements $u^{1,4}_{-\frac{3}{2}} = \frac{\theta E_{10}(4\tau)}{\Delta(4\tau)}$ and $v^{1,4}_{-\frac{3}{2}} = \frac{u^{1,0}_{\frac{5}{2}} E_8(4\tau)}{\Delta(4\tau)}$, which we combine into a combination $f^{1,4}_{-\frac{3}{2}} = (5u^{1,4}_{-\frac{3}{2}} + v^{1,4}_{-\frac{3}{2}})/6$ for which the third order pole $\frac{1}{q^3}$ vanishes. Further we consider $f^{1,3}_{-\frac{3}{2}} = -\frac{1}{4} \frac{[\theta, E_8(4\tau)]}{\Delta(4\tau)}$. Using the results of Borcherds and matching a finite number of genus zero invariants it can hence be proven that

$$\frac{2\Phi_{1,n}}{\eta^{24}}(q^4) = 2f^{1,4}_{-\frac{3}{2}} + nf^{1,3}_{-\frac{3}{2}}. \qquad (6.4.22)$$

For the fibration with $N = 2$ we use the general form (6.4.19). Then as before we construct $u^{2,0}_{\frac{5}{2}} = \theta E_2(8\tau) - 3\theta'(8\tau)$. In the next step we construct 4 functions of weight $-\frac{3}{2}$ with leading pole $\frac{1}{q^2}$: $u^{2,8}_{-\frac{3}{2}} = \frac{\theta E_{10}}{\Delta}$, $v^{2,8}_{-\frac{3}{2}} = \frac{u^{2,0}_{\frac{5}{2}} E_8}{\Delta}$, $w^{2,7}_{-\frac{3}{2}} = \frac{\theta E_8' - 16\theta' E_8}{\Delta}$ and $x^{2,7}_{-\frac{3}{2}} = \frac{u^{2,0}_{\frac{5}{2}} E_6' - 16\theta' E_6}{\Delta}$. One needs these in order to subtract all subleading poles and to define $f^{2,4}_{-\frac{3}{2}} = \frac{1}{576} w^{2,8}_{-\frac{3}{2}} + \frac{5}{864} x^{2,8}_{-\frac{3}{2}}$, $f^{2,7}_{-\frac{3}{2}} = (u^{2,8}_{-\frac{3}{2}} - v^{2,8}_{-\frac{3}{2}})/3 - 8f^{2,4}_{-\frac{3}{2}}$ and finally $f^{2,8}_{-\frac{3}{2}} = u^{2,8}_{-\frac{3}{2}} - f^{2,7}_{-\frac{3}{2}} - 2f^{2,4}_{-\frac{3}{2}}$. Now we find the result for

$$\frac{2\Phi_{2,n}}{\eta^{24}}(q^8) = 2f^{2,8}_{-\frac{3}{2}} + 2nf^{2,4}_{-\frac{3}{2}}. \qquad (6.4.23)$$

It is conceivable that there exists a more general family of regular $K3$ fibrations with the $N = 2$ fibre type, which involve $mf^{2,7}_{-\frac{3}{2}}$, but they have not been determined, yet (maybe one should check in Kreuzers list for the Euler numbers).

The advantage of the method is that it gives the answer for all regular one parameter families of $K3$ in a systematic manner. Since the construction should be clear by now we list here only the significant terms of the relevant $f^{3,n}_{-\frac{3}{2}}$ for the case with $N = 3$.

$$\begin{aligned}
f^{3,3}_{-\frac{3}{2}} &= q^{-3} - 2 + 6q - 12q^4 + 14q^9 + \\
f^{3,8}_{-\frac{3}{2}} &= q^{-8} - 27 + 56q + 214q^4 - 1512q^9 \\
f^{3,11}_{-\frac{3}{2}} &= q^{-11} - 54 - 134q + 924q^4 + 10098q^9 \\
f^{3,12}_{-\frac{3}{2}} &= q^{-12} - 74 - 336q - 2730q^4 - 17680q^9
\end{aligned} \qquad (6.4.24)$$

We find that

$$\frac{2\Phi_{1,n}}{\eta^{24}}(q^{12}) = 2f^{3,12}_{-\frac{3}{2}} - 4nf^{3,3}_{-\frac{3}{2}} \qquad (6.4.25)$$

reproduces the BPS numbers of the $N = 3$ fibrations. Examples with $n = 0$ are the complete intersection CY $X_{3,2}(1,1,1,1,1)$ over $\mathbb{P}^1$ as well as $X_{6,4}(1,1,2,2,2,2)$.

Let us finish the section with the description how to obtain from (6.4.8) the actual topological string partition functions of the topological string in the fibre direction. If $\frac{d^2}{2N} = l \in \mathbb{Z}$ we replace the power $\lambda^{2g-2} q^l \to \frac{\lambda^{2g-2}}{(2\pi i)^{3-2g}} Li_{3-2g}(e^{2\pi i d t})$, where t is the Kähler parameter of the fibre. All other q^r powers are dropped. With this information and the multicovering formula we get the following higher genus BPS invariants in the fibre direction.

6.4. PHYSICAL BOUNDARY CONDITIONS

$d=0$	1	2	3	4	5	6
g						
0	252	2496	223752	38637504	9100224984	2557481027520	805628041231176
1	4	0	-492	-1465984	-1042943520	-595277880960	-316194812546140
2	0	0	-6	7488	50181180	72485905344	70378651228338
3	0	0	0	0	-902328	-5359699200	-10869145571844
4	0	0	0	0	1164	228623232	1208179411278
5	0	0	0	0	12	-4527744	-94913775180
6	0	0	0	0	0	17472	4964693862
7	0	0	0	0	0	0	-152682820
8	0	0	0	0	0	0	2051118
9	0	0	0	0	0	0	-2124
10	0	0	0	0	0	0	-22

Table 6.4.1: BPS numbers for the $N=1$ fibre

$d=0$	1	2	3	4	5	6	7
g							
0	168	640	10032	288384	10979984	495269504	24945542832	1357991852672
1	4	0	0	-1280	-317864	-36571904	-3478901152	-306675842560
2	0	0	0	0	472	875392	220466160	36004989440
3	0	0	0	0	8	-2560	-6385824	-2538455296
4	0	0	0	0	0	0	50160	101090432
5	0	0	0	0	0	0	0	-1775104
6	0	0	0	0	0	0	0	4480

Table 6.4.2: BPS numbers for the $N=2$ fibre

6.4.3 The Seiberg-Witten plane

The emergence of the Seiberg-Witten plane as a divisor in the type II moduli space can be seen best in the case of the Calabi-Yau $\mathbb{P}_4^{(1,1,2,2,6)}$ [12]. In terms of the T and S moduli of the heterotic string one finds that the mirror map (6.3.18) can be written as [87] [82]

$$x = \frac{1728}{j(T)} + \cdots, \quad y = \exp(-S) + \cdots. \qquad (6.4.26)$$

This immediately translates to a powerful relation between the $SU(2)$ enhanced symmetry point of the heterotic model at $T=i$ and the singular point of the conifold locus on the type IIA side. Observe that $j(i) = 1728$ which inserted into the duality map (6.4.26) gives $x=1$. On the type II side this is the point of tangency between the conifold divisor C_{con} and the weak coupling divisor C_∞. Blowing up this singularity twice through inserting two $\mathbb{P}^1$'s gives the picture in figure 6.1. The divisor E_2 describes the physics of the Seiberg-Witten plane [11] for rigid $SU(2)$ Yang-Mills theory once we decouple gravity. In order to see this we will follow the analysis in [101]. As in the Seiberg-Witten theory the variable $u = \mathrm{tr}\phi^2$ vanishes at the $SU(2)$ point, it should be identified with $x-1$ to

leading order
$$x = 1 + \alpha' u + \mathcal{O}(\alpha'^2), \quad (6.4.27)$$
where the powers of α' are chosen such as to make the above expansion dimensionally correct. The second coordinate y is related to the $SU(2)$ scale Λ and coupling constant $e^{-\hat{S}}$ through
$$y = \alpha'^2 \Lambda^4 \exp(-\hat{S}) =: \epsilon^2. \quad (6.4.28)$$
The above identifications translate the conifold locus $(1-x)^2 - x^2 y = 0$ to
$$u^2 = \Lambda^4 \exp(-\hat{S}). \quad (6.4.29)$$

Decoupling gravity now means sending $\alpha' \to 0$. That is we construct the variables $x_1 = x^2 y/(x-1)^2$ and $x_2 = (x-1)$ which to leading order in α' correspond to $1/\tilde{u}^2$ and $\epsilon \tilde{u}$ [4]. These variables describe the Seiberg-Witten plane consistently at the semiclassical limit $\tilde{u} = \infty$, at the massless monopole points $\tilde{u} = \pm 1$ and at the $\mathbb{Z}_2$ orbifold point $\tilde{u} = 0$. It was shown in [101] that one obtains the rigid periods a, a_D as a subset of the periods of the Calabi-Yau by specialization of the Picard-Fuchs system to the semi-classical regime, $\tilde{u} \to \infty$:
$$(1, S, \sqrt{\alpha'} a, \sqrt{\alpha'} a_D, \alpha' u, \alpha' u S). \quad (6.4.30)$$

As at the monopole point a charged dyon gets massless we expect that in the limit where gravity becomes important this picture translates to a charged black hole becoming massless. Therefore, we expect that the topological amplitudes at this point will admit the conifold expansion (3.7.27).

6.4.4 The Gepner point

As Gepner found out [102], there is a Calabi-Yau minimal-model correspondence at the Fermat point in moduli space of the mirror. This is the point where $\phi = \psi = 0$ and corresponds via the mirror map (6.3.18) to the deep interior point in the moduli space of the Calabi-Yau M. In the picture 6.1 this is the point of intersection of the divisors C_0 and $D_{-1,0}$. The CFT description arising here is the tensor product of minimal models each at level P_i such that
$$\sum_{j=1}^{d+2} \frac{3P_j}{P_j + 2} = 3d, \quad (6.4.31)$$
where d is the complex dimension of the Calabi-Yau. This implies that D, the least common multiple of the $P_j + 2$, satisfies
$$D = \sum_{j=1}^{d+2} \frac{D}{P_j + 2}. \quad (6.4.32)$$

---
[4] Here, $\tilde{u} = u/(\Lambda^2 e^{-\hat{S}/2})$ is the correct dimensionless variable to use, see [101]

6.5. SOLUTION OF THE MODELS

Therefore, one can interpret the Calabi-Yau equation

$$\sum_{j=1}^{d+2} x_j^{P_j+2} = 0 \qquad (6.4.33)$$

in the weighted projective space $\mathbb{P}^{d+1}(\frac{D}{P_1+2}, \cdots, \frac{D}{P_{d+2}+2})$ as the superpotential of a Landau-Ginzburg theory with chiral superfields x_i [103](see [32] for a more rigorous description). Then the conformal field theory description arises as the infrared fixed point of this theory.

The impact on the topological free energies F^g is that these have to be regular in an expansion around the CFT point imposing boundary conditions on the holomorphic ambiguity and in particular on the constants a_I appearing in (3.7.26).

6.5 Solution of the Models

In this section we present the results of our calculations for the model $\mathbb{P}_4^{(1,1,2,2,2)}[8]$. The computations and the boundary behaviour of the model $\mathbb{P}_4^{(1,1,2,2,6)}$ is very similar and we therefore omit its discussion here and refer to reference [5] for further details.

6.5.1 $M_1 = \mathbb{P}_4^{(1,1,2,2,2)}[8]$

The toric data describing the ambient space of this Calabi-Yau can be summarized in the following table, where by V we denote the collection of vectors $\bar{v} = (v, 1)$ with v being the integral points of the polyhedron Δ^* and L are the corresponding charge vectors.

$$(V|L) = \begin{pmatrix} -1 & -2 & -2 & -2 & 1 & 0 & 1 \\ 1 & 0 & 0 & 0 & 1 & 0 & 1 \\ 0 & 1 & 0 & 0 & 1 & 1 & 0 \\ 0 & 0 & 1 & 0 & 1 & 1 & 0 \\ 0 & 0 & 0 & 1 & 1 & 1 & 0 \\ 0 & -1 & -1 & -1 & 1 & 1 & -2 \\ 0 & 0 & 0 & 0 & 1 & -4 & 0 \end{pmatrix} \qquad (6.5.1)$$

Here, the vector $v = (0, -1, -1, -1) = \frac{1}{2}(-1, -2, -2, -2) + \frac{1}{2}(1, 0, 0, 0)$ arises through the blow-up of the unique singularity in $\mathbb{P}_4^{(1,1,2,2,2)}$. Formula (6.1.3) gives $h^{1,1}(M_1) = 2$ and furthermore from (6.5.1) we can deduce the following quantities

(1) $\mathcal{D}_1 = \Theta_1^2(\Theta_1 - 2\Theta_2) - 4z_1(4\Theta_1 + 3)(4\Theta_1 + 2)(4\Theta_1 + 1)$
$\mathcal{D}_2 = \Theta_2^2 - z_2(2\Theta_2 - \Theta_1 + 1)(2\Theta_2 - \Theta_1)$
$\Delta_{con} = -1 + 512z_1 + 65536z_1^2(-1 + 4z_2)$
$\Delta_s = 1 - 4z_2$ \qquad (6.5.2)

(2) $\kappa_{111} = 8$, $\kappa_{112} = 4$, $\kappa_{222} = \kappa_{221} = 0$

(3) $\int_{M_1} c_2 J_1 = 56$, $\int_{M_1} c_2 J_2 = 24$

Solution at large radius

By the method of Frobenius we calculate the periods at large radius as solutions of the Picard-Fuchs system. This allows us to deduce the mirror map from (6.2.12)

$$\begin{aligned} 2\pi i t_1(z_1, z_2) &= \log(z_1) + 104 z_1 + 9780 z_1^2 - z_2 + 48 z_1 z_2 - \tfrac{3}{2} z_2^2 + \mathcal{O}(z^3), \\ 2\pi i t_2(z_1, z_2) &= \log(z_2) + 48 z_1 + 6408 z_1^2 + 2 z_2 - 96 z_1 z_2 + 3 z_2^2 + \mathcal{O}(z^3). \end{aligned} \quad (6.5.3)$$

These series can be inverted through introducing $q_i = e^{2\pi i t_i}$ and one obtains

$$\begin{aligned} z_1(q_1, q_2) &= q_1 - 104 q_1^2 + 6444 q_1^3 + q_1 q_2 - 304 q_1^2 q_2 + \mathcal{O}(q^4), \\ z_2(q_1, q_2) &= q_2 - 48 q_1 q_2 - 262 q_1^2 q_2 - 2 q_2^2 + 240 q_1 q_2^2 + 3 q_2^3 + \mathcal{O}(q^4). \end{aligned} \quad (6.5.4)$$

The Yukawa-couplings can be deduced from the Picard-Fuchs system as is explained in appendix A. Using the classical intersection numbers for normalization we obtain

$$C_{111} = \frac{-8}{z_1^3 \Delta_{con}}, \qquad C_{112} = \frac{4(256 z_1 - 1)}{z_1^2 z_2 \Delta_{con}},$$
$$C_{122} = \frac{8 - 4096 z_1}{z_1 z_2 \Delta_s \Delta_{con}}, \qquad C_{222} = \frac{-4(1 + 4 z_2 - 256 z_1(1 + 12 z_2))}{z_2^2 \Delta_s^2 \Delta_{con}}. \quad (6.5.5)$$

The Genus 0 invariants can be expressed in terms of these through the expansion

$$K_{ijk} = \frac{1}{X_0^2} C_{ijk}(t_1, t_2) = \partial_i \partial_j \partial_k F(t_1, t_2) = \kappa_{ijk} + \sum_{d_1, d_2} \frac{n_{d_1, d_2}^0 d_i d_j d_k}{1 - \prod_{l=1}^2 q_l^{d_l}} \prod_{l=1}^2 q_l^{d_l}. \quad (6.5.6)$$

In order to obtain the genus 1 free energy the holomorphic ambiguity has to be solved for. Using the ansatz (3.7.23) and as relevant boundary conditions $\int c_2 J_i$ as well as the known genus one GV invariants we arrive at

$$\mathcal{F}^{(1)} = \log\left(\Delta_{con}^{-\frac{1}{12}} \Delta_s^{-\frac{5}{12}} \exp\left[\frac{K}{2}(5 - \frac{\chi}{12})\right] \det G_{ij}^{-1} z_1^{-\frac{17}{6}} z_2^{-\frac{7}{8}} \right), \quad (6.5.7)$$

and sending $\bar{t} \to \infty$ the following holomorphic limit

$$\begin{aligned} F^1(t_1, t_2) &= -\frac{7}{3} \log(q_1) - \log(q_2) + \frac{160}{3} q_1 + \frac{2588}{3} q_1^2 + \frac{204928}{9} q_1^3 \\ &+ \frac{1}{3} q_2 + \frac{160}{3} q_1 q_2 + \frac{18056}{3} q_1^2 q_2 + \mathcal{O}(q^4). \end{aligned} \quad (6.5.8)$$

For higher genus calculations all propagators have to be obtained and $\mathcal{F}^{(1)}$ has to be brought into the form (3.7.6). Let us first concentrate on the propagators, they are determined through the equations (3.7.5) where the holomorphic ambiguities are solved for using symmetry properties. For S^{ij} independence on the index k and symmetry in the indices i and j is enough to fix all f_{lk}^i:

$$f_{11}^1 = -\frac{1}{z_1}, \; f_{12}^1 = -\frac{1}{4 z_2}, \; f_{22}^1 = 0,$$
$$f_{11}^2 = 0, \; f_{12}^2 = \frac{1}{2 z_1}, \; f_{22}^2 = -\frac{1}{z_2}. \quad (6.5.9)$$

6.5. SOLUTION OF THE MODELS

These choices lead to the following series expansions of the S^{ij}

$$S^{11} = -\frac{1}{16}z_1^2 + 16z_1^3 + \frac{1}{4}z_1^2 z_2 - 152z_1^3 z_2 + \mathcal{O}(z^5),$$

$$S^{12} = \frac{1}{8}z_1 z_2 - 38z_1^2 z_2 + 420z_1^3 z_2 - \frac{1}{2}z_1 z_2^2 + 152z_1^2 z_2^2 + \mathcal{O}(z^5),$$

$$S^{22} = -\frac{1}{4}z_2^2 + 144z_1^2 z_2^2 + z_2^3 + \mathcal{O}(z^5). \tag{6.5.10}$$

We find that the ambiguities appearing in the other propagators can be set to zero and obtain the following series expansions.

$$S^1 = \frac{3}{2}z_1^2 - 105z_1^3 - 12600z_1^4 + 6z_1^2 z_2 + 1548z_1^3 z_2 + \mathcal{O}(z^5),$$

$$S^2 = 3z_1 z_2 + 282z_1^2 z_2 + 33552z_1^3 z_2 - 12z_1 z_2^2 - 1128z_1^2 z_2^2 + \mathcal{O}(z^5),$$

$$S = -18z_1^2 - 2088z_1^3 - 320328z_1^4 - 144z_1^2 z_2 - 60480z_1^3 z_2 + \mathcal{O}(z^5). \tag{6.5.11}$$

Next, we turn our attention to the truncation relations (3.7.3). The ambiguities h_i^{jk}, h_j^i, h_i and h_{ij} are fixed through a series expansion of the holomorphic limit on both sides of (3.7.3)

$$h_1^{11} = -\frac{1}{32}z_1(1 - 4z_2 + 512z_1(11z_2 - 1)), \quad h_1^{12} = \frac{1}{16}(704z_1 - 1)z_2(4z_2 - 1),$$

$$h_1^{22} = \frac{z_2^2(4z_2 - 1)}{8z_1},$$

$$h_2^{11} = \frac{z_1^2(1 + 4z_2 - 256z_1(1 + 15z_2))}{64z_2}, \quad h_2^{12} = \frac{1}{32}z_1(256z_1 - 1)(1 + 4z_2),$$

$$h_2^{22} = \frac{1}{16}z_2(1 + 4z_2 + 448z_1(4z_2 - 1)),$$

$$h_1^1 = 12z_1 z_2, \quad h_1^2 = 6(1 - 4z_2)z_2, \quad h_2^1 = 6z_1^2, \quad h_2^2 = -12z_1 z_2, \tag{6.5.12}$$

where all other ambiguities are either zero or follow by symmetry. Having obtained the truncation relations the process of direct integration demands for the covariant derivative of $\mathcal{F}^{(1)}$ in terms of the propagators,

$$D_i \mathcal{F}^{(1)} = \frac{1}{2}C_{ijk}S^{jk} - \frac{1}{12}\Delta_{con}^{-1}\partial_i \Delta_{con} - \frac{5}{12}\Delta_s^{-1}\partial_i \Delta_s + \partial_i \log(z_1^{-\frac{31}{12}} z_2^{-\frac{7}{8}}). \tag{6.5.13}$$

From here it is now straightforward to carry out the integration. First make an ansatz for $\mathcal{F}^{(2)}$ as a polynomial of degree $3g - 3$ in the generators $\tilde{S}^{ij}$, $\tilde{S}^i$ and $\tilde{S}$ and evaluate the right hand side of the first equation in (3.7.8) by applying covariant derivatives to $D_i \mathcal{F}^{(1)}$ and using the truncation relations (3.7.3). The right coefficients in $\mathcal{F}^{(2)}$ follow then from comparison. In order to apply this procedure iteratively genus by genus the holomorphic ambiguity f_g has to be fixed at each step. This is done as discussed in section (3.7.3) by going to various boundary divisors in moduli space as will be described in the succeeding paragraphs.

Solution at the Conifold locus

We choose the monopole point of the Seiberg-Witten plane in order to carry out the conifold expansion. As was noted in section (6.4.3) the correct coordinates at this point are

$$z_{c,1} = \frac{x^2 y}{(x-1)^2} - 1, \quad z_{c,2} = x - 1. \tag{6.5.14}$$

Transforming the Picard-Fuchs system to these coordinates we find the following solutions (see also [104])

$$\begin{aligned}
\omega_0^c &= 1 + z_{c,2} - \frac{73}{64}z_{c,2}^2 - \frac{73}{192}z_{c,1}z_{c,2}^2 + \frac{7043}{4608}z_{c,2}^3 + \mathcal{O}(z_c^4), \\
\omega_1^c &= z_{c,1}\sqrt{z_{c,2}} - \frac{15}{32}z_{c,1}^2\sqrt{z_{c,2}} + \frac{315}{1024}z_{c,1}^3\sqrt{z_{c,2}} + \frac{11}{64}z_{c,1}^2 z_{c,2}^{3/2} + \mathcal{O}(z_c^5), \\
\omega_2^c &= z_{c,2} - \frac{35}{32}z_{c,2}^2 - \frac{35}{96}z_{c,1}z_{c,2}^2 + \frac{3325}{2304}z_{c,2}^3 + \mathcal{O}(z_c^4),
\end{aligned} \tag{6.5.15}$$

where we have suppressed the dual logarithmic solutions. As mirror coordinates we take $t_{c,1} := \frac{\omega_1^c}{\omega_0^c}$ and $t_{c,2} := \frac{\omega_2^c}{\omega_0^c}$. Relevant for the expansions around $z_{c,1} = z_{c,2} = 0$ is the inverse mirror map given by

$$\begin{aligned}
z_{c,1}(t_{c,1}, t_{c,2}) &= \frac{t_{c,1}}{\sqrt{t_{c,2}}} + \frac{15}{32}\frac{t_{c,1}^2}{t_{c,2}} - \frac{3}{64}t_{c,1}\sqrt{t_{c,2}} + \frac{135}{1024}\frac{t_{c,1}^3}{t_{c,2}^{3/2}} + \frac{1515}{65536}\frac{t_{c,1}^4}{t_{c,2}^2} - \frac{1223}{3072}t_{c,1}^2 + \mathcal{O}(t_c^3), \\
z_{c,2}(t_{c,1}, t_{c,2}) &= t_{c,2} + \frac{67}{32}t_{c,2}^2 + \frac{35}{96}t_{c,1}t_{c,2}^{3/2} + \frac{175}{1024}t_{c,1}^2 t_{c,2} + \frac{18847}{4608}t_{c,2}^3 + \mathcal{O}(t_c^4).
\end{aligned} \tag{6.5.16}$$

In order to perform the above inversion one has to define new variables $s_1 = \frac{t_{c,1}}{\sqrt{t_{c,2}}}$ and $s_2 = t_{c,2}$, calculate $z_{c,i}(s_1, s_2)$ and then insert back $s_i(t_{c,i})$. The divisor $\{z_{c,2} = 0\}$ is normal to the conifold locus and $\{z_{c,1} = 0\}$ is tangential. This means that $z_{c,1}$ parameterizes the normal direction to the conifold locus and $z_{c,2}$ the tangential one. Therefore we expect $t_{c,1}$ to be appearing in inverse powers in the expansion of the free energies.

To obtain the free energies, all nonholomorphic generators appearing in the polynomial expansion of the $\mathcal{F}^{(g)}$ have to be transformed into the $z_{c,i}$ coordinates. That is Yukawa couplings, the Christoffel symbols and the holomorphic ambiguities f appearing in (3.7.5) must be transformed to the conifold coordinates. For the f_{jk}^i this means

$$f_{jk}^i(z_c) = \frac{\partial z_{c,i}}{\partial z_l}\left(\frac{\partial^2 z_l}{\partial z_{c,j}\partial z_{c,k}}\right) + \frac{\partial z_{ci}}{\partial z_l}\frac{\partial z_m}{\partial z_{c,j}}\frac{\partial z_n}{\partial z_{c,k}}f_{mn}^l(z), \tag{6.5.17}$$

while the Yukawa-couplings have to be tensor transformed and the Christoffel symbols are obtained directly from the periods (6.5.15).

6.5. SOLUTION OF THE MODELS

We display our results for the F_c^g up to genus 3:

$$F_c^1 = -\frac{1}{12}\log\left(\frac{t_{c,1}}{\sqrt{t_{c,2}}}\right) - \frac{29}{12}\log(t_{c,2}) - \frac{137}{128}\frac{t_{c,1}}{\sqrt{t_{c,2}}} - \frac{9827}{768}t_{c,2} + \frac{189}{8192}\frac{t_{c,1}^2}{t_{c,2}} + \mathcal{O}(t_c^2),$$

$$F_c^2 = -\frac{1}{240t_{c,1}^2} + \frac{155359}{589824} - \frac{550551 t_{c,1}^2}{33554432 t_{c,2}^2} - \frac{18321 t_{c,1}}{524288 t_{c,2}^{3/2}} + \frac{1067}{6144 t_{c,2}} + \mathcal{O}(t_c^1),$$

$$F_c^3 = \frac{1}{1008 t_{c,1}^4} + \frac{788437361}{21743271936} + \mathcal{O}(t_c^1) \tag{6.5.18}$$

Solution at the strong coupling locus

We expand around the point of intersection of the divisors C_1 and $D_{0,-1}$. The right coordinates are

$$z_{s,1} = \Delta_s^{\frac{1}{2}}, \; z_{s,2} = x_1. \tag{6.5.19}$$

The Picard-Fuchs system at this point contains among its solutions a one logarithmic in x_1 which is a continuation of the logarithmic solution already present at large radius.

$$\omega_0^s = 1 + \frac{3}{32}z_{s,2} + \frac{945}{16384}z_2^2 + \frac{28875}{524288}z_2^3 + \mathcal{O}(z_s^4),$$

$$\omega_1^s = z_{s,1} + \frac{z_{s,1}^3}{3} + \frac{1}{32}z_{s,1}^3 z_{s,2} + \mathcal{O}(z_s^5),$$

$$\omega_2^s = \omega_0^s \log(z_{s,2}) - \frac{1}{2}z_{s,1}^2 + \frac{z_{s,2}}{2} + \frac{2853}{8192}z_{s,2}^2 - \frac{3}{64}z_{s,1}^2 z_{s,2}$$

$$+ \frac{273425}{786432}z_{s,2}^3 + \mathcal{O}(z_s^4). \tag{6.5.20}$$

The mirror map is deduced from the quotients by ω_0^s: $t_{s,1} := \frac{\omega_1^s}{\omega_0^s}$, $t_{s,2} := \frac{\omega_2^s}{\omega_0^s}$. Building the inverse we arrive at

$$z_{s,1}(t_{s,1}, q_{s,2}) = t_{s,1} + \frac{3}{32}q_{s,2}t_{s,1} + \frac{177}{16384}q_{s,2}^2 t_{s,1} - \frac{t_{s,1}^3}{3} + \mathcal{O}(t_c^4),$$

$$z_{s,2}(t_{s,1}, q_{s,2}) = q_{s,2} - \frac{q_{s,2}^2}{2} + \frac{603}{8192}q_{s,2}^3 + \frac{1}{2}q_{s,2}t_{s,1}^2 + \mathcal{O}(t_c^4). \tag{6.5.21}$$

Transforming the Yukawa couplings, the Christoffel symbols and the holomorphic ambiguities f to the strong coupling coordinates we obtain the propagators at this point. Tensor transforming the propagators to z_i coordinates and substituting in all holomorphic quantities in the $\mathcal{F}^{(g)}$ $z_i \to z_i(z_{s,1}, z_{s,2})$ we obtain the $\mathcal{F}_s^{(g)}$. In the holomorphic limit this gives the following genus 1,2 and 3 expansions.

$$\begin{aligned}
F_s^1 &= -\frac{1}{3}\log(t_{s,1}) - \frac{7}{3}t_{s,2} + \frac{5}{24}q_{s,2} + \frac{121}{4096}q_{s,2}^2 - \frac{t_{s,1}}{18} + \frac{5}{48}q_{s,2}t_{s,1} \\
&\quad + \frac{t_{s,2}^2}{540} + \mathcal{O}(t_s^3), \\
F_s^2 &= \frac{1}{240t_{s,1}^2} - \frac{11}{360} + \frac{q_{s,2}}{1280} - \frac{1057}{2097152}\frac{q_{s,2}^3}{t_{s,1}^2} + \mathcal{O}(t_s^2), \\
F_s^3 &= \frac{1}{4032t_{s,1}^4} + \frac{11}{90720} - \frac{q_{s,2}}{16128} + \frac{18805}{132120576}\frac{q_{s,1}^3}{t_{s,1}^2} \\
&\quad - \frac{62210349}{549755813888}\frac{q_{s,2}^5}{t_{s,1}^4} + \mathcal{O}(t_s^2).
\end{aligned} \qquad (6.5.22)$$

It is important to note here that the variable $t_{s,1}$ is not the true variable characterizing the size of the shrinking cycle. The true size is given by the rescaling $t_{s,1} \to 2t_{s,1}$. The factor of 4 which then arises in the expansions in comparison to the conifold case is exactly the difference between hyper- and vector multiplets calculated in section (6.4.1). However, note that we have only displayed results up to genus 3. For genus 4 we find that there is no simple gap structure and therefore the boundary conditions at this point remain unclear. Also, we find that for genus 4 the parameterization of the ambiguity changes slightly as the numerator of the strong coupling discriminant has to be parametrized to a higher degree

$$f_g = \ldots + \frac{\sum_{|I|\leq 2g-2} c_I^s z^I}{\Delta_s^{g-1}} + \ldots. \qquad (6.5.23)$$

As in this modified ansatz some of the parameters lead to contributions of non regular terms at the Gepner point they will be solved for by the regularity condition at that point.

The strong coupling divisor provides us with a further boundary condition to check the consistency of our approach. At this locus there is an extremal transition to the 1-parameter complete intersection $\mathbb{P}^5[4,2]$. The free energy expansions for $\mathbb{P}^5[4,2]$ should come out naturally from the expansion at large radius of our 2-parameter model by setting $q_2 = 1$ which corresponds to shrinking the size of the $\mathbb{P}^1$ base to zero [5]. Therefore the Gopakumar-Vafa invariants of $\mathbb{P}^5[4,2]$ should be equal to the sum over the second degree of the invariants of M_1, i.e. $n_k^{(g)} = \sum_{d_2} n_{d_1 d_2}^{(g)}$. This is indeed true as can be checked from the tables in appendix E.1.

---

[5]In the general case the transition is obtained by replacing $(2g-2)\binom{N}{2}$ two spheres by $(2g-2)\binom{N}{2}$ three spheres, where g is the genus of the singular curve and $N-1$ is the rank of the gauge group. This results in a change of Hodge numbers given by $h^{1,1} \mapsto h^{1,1} - (N-1)$, $h^{2,1} \mapsto h^{2,1} + (2g-2)\binom{N}{2} - (N-1)$.

Solution at the Gepner point

Here we expand around the point of intersection of the divisors C_0 and $D_{-1,0}$ which corresponds to $\psi = \phi = 0$. This gives the variables

$$z_{o,1} = \psi = \frac{1}{(z_2 z_1^2)^{\frac{1}{8}}}, \quad z_{o,2} = \phi = \frac{1}{z_2^{\frac{1}{2}}}, \tag{6.5.24}$$

where we have chosen the subscript o as the Gepner point is a Landau-Ginzburg orbifold. The transformed Picard-Fuchs system admits the solutions

$$\begin{aligned}
\omega_0^o &= z_{o,1} + \frac{1}{32}z_{o,1}z_{o,2}^2 + \frac{27}{2048}z_{o,1}z_{o,2}^4 + \mathcal{O}(z_o^6), \\
\omega_1^o &= z_{o,1}z_{o,2} + \frac{25}{96}z_{o,1}z_{o,2}^3 + \frac{z_{o,1}^5}{6} + \mathcal{O}(z_o^6), \\
\omega_2^o &= z_{o,1}^2 + \frac{1}{8}z_{o,1}^2 z_{o,2}^2 + \mathcal{O}(z_o^6), \\
\omega_3^o &= z_{o,1}^2 z_{o,2} + \frac{3}{8}z_{o,1}^2 z_{o,2}^3 + \mathcal{O}(z_o^6), \\
\omega_4^o &= z_{o,1}^3 + \frac{9}{32}z_{o,1}^3 z_{o,2}^2 + \mathcal{O}(z_o^6),
\end{aligned} \tag{6.5.25}$$

where we have omitted the solution corresponding to the 6th period. From the above we extract the mirror map $t_{1,o} := \frac{\omega_1^o}{\omega_0^o}$, $t_{2,o} := \frac{\omega_2^o}{\omega_0^o}$ and its inverse

$$\begin{aligned}
z_{o,1}(t_{o,1}, t_{o,2}) &= t_{o,1} - \frac{3}{32}t_{o,1}t_{o,2}^2 + \frac{17}{6144}t_{o,1}t_{o,2}^4 + \mathcal{O}(t_o^6), \\
z_{o,2}(t_{o,1}, t_{o,2}) &= t_{o,2} - \frac{11}{48}t_{o,2}^3 - \frac{1}{6}t_{o,1}^4 + \frac{31}{768}t_{o,2}^5 + \mathcal{O}(t_o^6).
\end{aligned} \tag{6.5.26}$$

The genus 0 Prepotential can be extracted from the periods ω_3^o and ω_4^o through the special geometry relation

$$\omega_3^o = \omega_0^o \left(\frac{\partial}{\partial t_{o,1}}F_o(t_{o,1}, t_{o,2})\right), \quad \omega_4^o = \omega_0^o \left(\frac{\partial}{\partial t_{o,2}}F_o(t_{o,1}, t_{o,2})\right), \tag{6.5.27}$$

yielding

$$F_o(t_{o,1}, t_{o,2}) = t_{o,1}^2 t_{o,2} + \frac{1}{48}t_{o,1}^2 t_{o,2}^3 + \frac{t_{o,1}^6}{30} + \mathcal{O}(t_o^7). \tag{6.5.28}$$

The $F_o^g(t_{o,1}, t_{o,2})$ can be calculated in the same way as in the case of the other boundary divisors and we find that as expected the correct GV-invariants are produced once we require the free energies to be regular at the orbifold point. This way also all polynomial ambiguities a_I are fixed uniquely. Let us delve a bit more into the details at this point as the question of fixing the a_I is ultimately related to the question of integrability. We observe that the anomaly part of the F_o^g always comes with a pole $\frac{1}{z_{o,1}^{2g-2}} + \ldots$ and that no poles of type $\frac{1}{z_{o,2}}$ appear. This can be traced back to the expansion of propagators and

Yukawa couplings around this point. Therefore, we are able to constrain the parameterization of the ambiguity in the following way: The coefficients a_{i_1,i_2} will be nonvanishing only for indices $i_2 \leq i_1/2$. The reason can be found in the relations inverse to (6.5.24)

$$z_1 = -\frac{z_{o,2}}{256 z_{o,4}^4}, \quad z_2 = \frac{1}{4 z_{o,2}^2}. \tag{6.5.29}$$

One can see that in order to avoid poles in the second variable $z_{o,2}$ one has to multiply each power of z_2 with a power of z_1 which is at least two times larger. We find that at genus 2 the regularity condition is enough to fix all a_I's without exception. On the other hand the calculation at genus 3 yields that regularity is only strong enough to fix all but one of the a_I, namely $a_{1,0}$. However, this single parameter is already solved for by the knowledge of the fiber invariants extracted from the weak coupling divisor. At genus 4 we find that two parameters remain unfixed after having imposed regularity, namely $a_{1,0}$ and $a_{2,0}$. Again these two will be ultimately solved for by the knowledge of the fibre invariants once all other parameters are fixed. This procedure will carry on up to genus infinity once the parameters related to the strong coupling divisor and the conifold divisor can be fixed at each genus by appropriate boundary conditions.

We display our results for the higher $F_o^g(t_{o,1}, t_{o,2})$ $(g \geq 1)$:

$$\begin{aligned}
F_o^1(t_{o,1},t_{o,2}) &= -\frac{3}{16}t_{o,2}^2 + \frac{7}{384}t_{o,2}^4 - \frac{1}{16}t_{o,1}^4 t_{o,2} + \frac{43}{184320}t_{o,2}^6 - \frac{5}{256}t_{o,1}^4 t_{o,2}^3 \\
&\quad -\frac{5}{252}t_{o,1}^8 + \frac{2237}{20643840}t_{o,2}^8 - \frac{35}{6144}t_{o,1}^4 t_{o,2}^5 - \frac{7}{1280}t_{o,1}^8 t_{o,2}^2 \\
&\quad +\frac{40603}{2972712960}t_{o,2}^{10} + \mathcal{O}(t_o^{11}), \\
F_o^2(t_{o,1},t_{o,2}) &= \frac{113}{7680}t_{o,1}^2 - \frac{377}{122880}t_{o,1}^2 t_{o,2}^2 + \frac{363}{655360}t_{o,1}^2 t_{o,2}^4 \\
&\quad +\frac{59}{307200}t_{o,1}^6 t_{o,2} - \frac{153361}{707788800}t_{o,1}^2 t_{o,2}^6 + \frac{39041}{44236800}t_{o,1}^6 t_{o,2}^3 \\
&\quad -\frac{143}{3440640}t_{o,1}^{10} - \frac{10379101}{158544691200}t_{o,1}^2 t_{o,2}^8 + \mathcal{O}(t_o^{11}),
\end{aligned}$$

$$\begin{aligned}
F_o^3(t_{o,1},t_{o,2}) &= -\frac{61}{3932160}t_{o,2} + \frac{29}{9437184}t_{o,2}^3 + \frac{875}{4718592}t_{o,1}^4 - \frac{581}{603979776}t_{o,2}^5 \\
&\quad +\frac{1585}{44040192}t_{o,1}^4 t_{o,2}^2 - \frac{236533}{1014686023680}t_{o,2}^7 + \frac{1221673}{12683575296}t_{o,1}^4 t_{o,2}^4 \\
&\quad -\frac{43439}{1056964608}t_{o,1}^8 t_{o,2} - \frac{6903751}{73057393704960}t_{o,2}^9 + \frac{20689415}{304405807104}t_{o,1}^4 t_{o,2}^6 \\
&\quad +\mathcal{O}(t_o^{11}). \tag{6.5.30}
\end{aligned}$$

Chapter 7

Conclusions

In this thesis, after having pointed out the main physical principles and applications of topological string theory, we gave a detailed account on the construction of topological field theories and their coupling to gravity, followed by a thorough discussion of solutions to the holomorphic anomaly equations on the main three kinds of Calabi-Yau manifolds. From the point of view of the gauged linear σ model presented in section 3.2.2 these are the ones with a nonlinear gauge group, the ones with a linear gauge group but without superpotential, and finally the ones with a linear gauge group and a superpotential.

Our work and in particular the tables of BPS invariants presented in the appendices represent a test of mirror symmetry for situations where a mathematical proof is lacking and which are not well understood from the mathematics point of view. This includes the case of the Grassmannian Calabi-Yau manifolds for genus 0 and higher genera, and higher genus calculations ($g \geq 1$) for toric Calabi-Yau manifolds. This is due to the fact that the proof that mathematicians [105] gave for the fact that the B-model calculation of genus zero amplitude counts worldsheet instantons on the mirror relies on localization w.r.t. the $U(1)^r$ action. It has not been extended to the non-abelian case, nor to higher genera.

From the physics point of view the BPS invariants we have computed are important for the calculation of microscopic black hole entropies in the case of compact Calabi-Yau manifolds. We have performed tests of the equality of microscopic and macroscopic entropy for the Grassmannian Calabi-Yau models and found a good matching between the two results. This provides further support for the correct microscopic interpretation of the entropy of extremal five dimensional spinning black holes. In the case of K3 fibrations the BPS states also encode nonperturbative corrections to the Heterotic string, namely they count the number of gauge instantons as the exponential of the size of the $\mathbb{P}^1$-base gets mapped to e^{1/g_s^2}. Furthermore, it would be interesting to identify a CFT point for Calabi-Yau manifolds which come as hypersurfaces or complete intersections in Grassmannians. At a CFT point in moduli space the complete string theoretic spectrum of the model becomes computable and not only the massless one. We find that the model $(\mathbb{G}(2,5)\|1,1,3)^1_{-150}$ is regular at $t_\infty = 0$ at least to genus 5. This hints at a CFT point with a $\mathbb{Z}_3$ automorphism group. One possibility would be that the CFT description emerging there is equivalent to the coset models of Suzuki and Kazama [106].

One of the main objectives of our approach was the question of integrability of the topological string on Calabi-Yau backgrounds. For local Calabi-Yau geometries we find that the gap condition at the conifold is strong enough to fix all holomorphic ambiguity parameters. This makes the topological string integrable on non-compact Calabi-Yau manifolds whose mirror geometry is encoded by a family of Riemann surfaces Σ_g and a meromorphic differential λ. Indeed the successful direct integration in the case of non-compact models can be traced back to well known transformation properties of the topological string amplitudes w.r.t to the modular group of Σ_g, which is a finite index subgroup of $Sp(2g, \mathbb{Z})$. Moreover it can be established [107, 108] that the theory becomes equivalent to a matrix model whose eigenvalue dynamics and correlators can be completely fixed by its spectral curve and the meromorphic differential defining the filling fraction. These data are precisely identified with the Riemann surface Σ_g and λ respectively [107, 108].

In the case of K3 fibrations discussed in section 6 the situation is more involved. First of all the moduli space is much richer and apart from the conifold divisor the physical boundary conditions at several other divisors become important when fixing the holomorphic ambiguity. In mathematical terms the reason is that in the case of compact Calabi-Yau manifolds the modular groups are not well understood. Even for the Quintic one does not know whether the modular group is a finite index subgroup of $Sp(4, \mathbb{Z})$ or not. However, near the weak coupling divisor such a modular description becomes available which solves the theory in the fibre direction up to genus infinity. This suggests that augmenting the results at the weak coupling divisor with physical boundary conditions at other divisors might ultimately lead to the integrability of the topological string on K3 fibrations. Indeed we observe that the gap at the conifold, the regularity at the CFT point and a further gap structure at the strong coupling divisor fix the ambiguity completely up to genus 3. Surprisingly, at genus 4 we find that there is no simple gap structure at the strong coupling divisor any longer. The reason must lie in the gauge theory emerging at this divisor. As the divisor is of codimension one there will be an $SU(2)$ gauge enhancement at its locus in moduli space and the gauge theory will be in the Coulomb branch. This means that near the singularity at high energies there will be two light vector bosons, namely the W^+ and W^- bosons, and one massless $U(1)$. Apparently, this is an interacting gauge theory, so the question is not why the gap disappears at genus 4, but rather why it is present at genus 2 and 3. The answer must lie in the light spectrum of the effective field theory arising from the gauge theory at low energies. It might well be that the nontrivial cancellations which are present at genus 2 and 3 will also be there at higher genera leading to the classical gap structure (3.7.28). So, from the physics point of view, the question of integrability of the topological string on K3 fibrations seems to be ultimately related to the details of the low energy gauge theory arising at the strong coupling divisor.

Looking at topological string amplitudes on general Calabi-Yau manifolds and taking a more mathematical point of view we find that the monodromy behaviour of the periods at boundary divisors in moduli space has a direct impact on the structure of the amplitudes.

In particular one finds that the topological string amplitudes are regular at points of finite monodromy (examples are the maximal unipotent monodromy point and the Gepner point), and suffer from a gap like or singular behaviour at points where the monodromy is not of finite order[1]. Therefore, it would be interesting to determine more specific and stringent conditions on the monodromy at points where a full gap in the amplitudes is observed. Going further along this line of thought there is the hope that one can find out more about the automorphic forms, which the topological free energies represent, by analyzing their expansions around singular divisors in moduli space and classifying the monodromy around these. Yet, another way to set up a proof of integrability is to *patch* together local matrix model descriptions along a compact Calabi-Yau. We know, for example, that the conifold locally admits such a universal matrix model description. One could well succeed in finding such local descriptions on *charts* of the compact space, which can be glued together by nontrivial coordinate transformations[2].

One of the main reasons why it would be interesting to establish integrability on specific compact spaces such as K3 fibrations, is that these are related to different Calabi-Yau spaces, i.e. different *string vacua*, through extremal transitions [1]. Thus proving integrability for one class of spaces would shed light on the physics for a huge class of other models. A direct physical application would be the computation of the microscopic black hole entropy to very high accuracy and thus clarify issues related to the OSV-conjecture [2]. Another related important application would be the analysis of the nonperturbative completion of topological string theory. As one finds, topological string amplitudes are not convergent or Borel summable when going to high genus. The reason is that they constitute a perturbative series in λ, the graviphoton vev, and miss terms which come exponentially in λ. In a full description such terms might arise from KK-monopoles[3], which, however, is the subject of future work.

---

[1] I would like to thank Emanuel Scheidegger for pointing this out to me.
[2] Here, I have benefited greatly from discussions with Emanuel Scheidegger and Alireza Tavanfar.
[3] I would like to thank Stefan Vandoren for pointing this out to me.

Appendix A
Yukawa-couplings from Picard-Fuchs operators

In this appendix we present a method for obtaining Yukawa couplings from Picard-Fuchs equations following reference [78]. As explained in 3.1.2 Yukawa-couplings are functions of the complex structure moduli and are defined as

$$C_{ijk}(\underline{z}) = \int \Omega(\underline{z}) \wedge \partial_{z_i}\partial_{z_j}\partial_{z_k}\Omega(\underline{z})$$

$$= \sum_{l=0}^{h^{2,1}}(X^l \partial_{z_i}\partial_{z_j}\partial_{z_k}\mathcal{F}_l - \mathcal{F}_l \partial_{z_i}\partial_{z_j}\partial_{z_k}X^l). \quad \text{(A.0.1)}$$

We can now define

$$W^{(k_1,\cdots,k_d)} = \sum_l (X^l \partial_{z_1}^{k_1} \cdots \partial_{z_d}^{k_d} \mathcal{F}_l - \mathcal{F}_l \partial_{z_1}^{k_1} \cdots \partial_{z_d}^{k_d} X^l)$$

$$:= \sum_l (X^l \partial^{\mathbf{k}} \mathcal{F}_l - \mathcal{F}_l \partial^{\mathbf{k}} X^l), \quad \text{(A.0.2)}$$

where $1 \leq d \leq h^{2,1}$. We see that the various types of Yukawa couplings are described through $W^{\mathbf{k}}$ with $\sum k_i = 3$ and that $W^{\mathbf{k}} = 0$ for $\sum k_i = 0, 1, 2^1$. Next, write the Picard-Fuchs differential operators in the form

$$\mathcal{D}_l = \sum_{\mathbf{k}} f_l^{\mathbf{k}} \partial^{\mathbf{k}}, \quad \text{(A.0.3)}$$

which together with (A.0.2) gives

$$\sum_{\mathbf{k}} f_l^{\mathbf{k}} W^{(\mathbf{k})} = 0. \quad \text{(A.0.4)}$$

These differential equations can be supplemented by further operators obtained by applying further derivatives to the Picard-Fuchs operators, i.e. operators of the form

---
[1]This is due to the fact that second and lower order derivatives of Ω do not contain a $(0,3)$ part.

156 CHAPTER A. YUKAWA-COUPLINGS FROM PICARD-FUCHS OPERATORS

$\partial_{z_i}\mathcal{D}_l$. In the case of compact Calabi-Yau manifolds the differential equations obtained this way are complete, and can thus be integrated to give the Yukawa couplings up to an overall normalization. In order to perform the integration one has to make use of the following identities

$$\begin{aligned} W^{(4,0)} &= 2\partial_{z_1}W^{(3,0)} \\ W^{(3,1)} &= \frac{3}{2}\partial_{z_1}W^{(2,1)} + \frac{1}{2}\partial_{z_2}W^{(3,0)} \\ W^{(2,2)} &= \partial_{z_1}W^{(1,2)} + \partial_{z_2}W^{(2,1)}, \end{aligned} \qquad (A.0.5)$$

where we have specified to the two-parameter case as in this thesis we only look at models which have at most two complex structure deformations. However, integrating equations (A.0.4) is not yet the end of the story. An important task left is to fix the overall normalization which is done by transforming the Yukawa-couplings to their A model counterparts

$$K_{ijk} = \frac{1}{X_0^2}C_{ijk}(t_1,t_2) = \partial_i\partial_j\partial_k F^0(t_1,t_2) = \kappa_{ijk} + \sum_{d_1,d_2}\frac{n^0_{d_1,d_2}d_id_jd_k}{1-\prod_{l=1}^2 q_l^{d_l}}\prod_{l=1}^2 q_l^{d_l}, \qquad (A.0.6)$$

and solving for the normalization factor in terms of the classical intersection numbers κ_{ijk} of the manifold M [2].

---

[2]The expressions $C_{ijk}(t_1,t_2)$ are obtained by tensor transformation of the quantities $C_{ijk}(z_1(\underline{t}),z_2(\underline{t}))$.

Appendix B

Modular anomaly versus holomorphic anomaly

Physically the amplitudes F^g of the topological string are invariant under the space-time modular group Γ of the target space. This is the most important restriction on these functions. The nicest case is when the B-model geometry is a family of elliptic curves. Then Γ is a subgroup of $\mathrm{SL}(2,\mathbb{Z})$ and the classical theory of modular forms applies. We will recapitulate below the relevant aspects of $\mathrm{SL}(2,\mathbb{Z})$ almost holomorphic modular forms. This gives some insight in the interplay between the breaking of the modularity and the breaking of holomorphicity. The different modular forms that we need for the general families of elliptic curves, i.e. general two cut matrix models, follow from the Picard-Fuchs equations. The relation between the Picard-Fuchs equations and modular forms is again a classical subject, which has been beautifully reviewed in [52].

B.1 PSL(2, $\mathbb{Z}$) modular forms

We define $q := e^{2\pi i \tau}$, with $\tau \in \mathbb{H}_+ = \{\tau \in \mathbb{C} \,|\, \text{Im}(\tau) = \frac{1}{2i}(\tau - \bar{\tau}) > 0\}$ and the projective action PSL(2, $\mathbb{Z}$) of $\Gamma_1 = \text{SL}(2, \mathbb{Z}) = \left\{ \gamma = \begin{pmatrix} a & b \\ c & d \end{pmatrix} \,\Big|\, ad - bc = 1,\ a, b, c, d \in \mathbb{Z} \right\}$ on $\mathbb{H}_+$ by

$$\tau \mapsto \tau_\gamma = \frac{a\tau + b}{c\tau + d}, \tag{B.1.1}$$

for $\gamma \in \Gamma_1$. It follows that

$$\frac{1}{\text{Im}(\tau_\gamma)} = \frac{(c\tau + d)^2}{\text{Im}(\tau)} - 2ic(c\tau + d) = \frac{|c\tau + d|^2}{\text{Im}(\tau)}. \tag{B.1.2}$$

Modular forms of Γ_1 transform as

$$f_k(\tau_\gamma) = (c\tau + d)^k f_k(\tau) \tag{B.1.3}$$

with weight $k \in \mathbb{Z}$ for all $\tau \in \mathbb{H}_+$ and $\gamma \in \Gamma_1$, are meromorphic for $\tau \in \mathbb{H}_+$ and grow like $\mathcal{O}(e^{C \text{Im}(\tau)})$ for $\text{Im}(\tau) \to \infty$ and $\mathcal{O}(e^{C/\text{Im}(\tau)})$ for $\text{Im}(\tau) \to 0$ with $C > 0$. A strategy to build modular forms of weight k is to sum over orbits of Γ_1

$$G_k = \frac{1}{2} \sum_{\substack{m, n \in \mathbb{Z} \\ (m, n) \neq (0, 0)}} \frac{1}{(m\tau + n)^k}. \tag{B.1.4}$$

It is easy to see that this expression transforms like (B.1.3), converges absolutely for $k > 2$ and vanishes for k odd. In the standard definition of the Eisenstein series E_k the sum runs over coprime (m, n), which yields a proportionality $G_k(\tau) = \zeta(k) E_k(\tau)$, where $\zeta(k) = \sum_{n \geq 1} \frac{1}{n^k}$. One shows ([52]) the central fact that E_4, E_6 (or G_4, G_6 of course) generate freely the graded (by k) ring of modular forms $\mathcal{M}_*(\Gamma_1)$.

Still one may spot two shortcomings. Firstly the ring $\mathcal{M}_*(\Gamma_1)$ does not close under any differentiation and secondly there should be a modular form for weight 2. These facts are related as $d_\tau = \frac{d}{2\pi i d\tau}$ has weight 2. The second is remedied by an ϵ regularization in the sum $G_{2,\epsilon} = \frac{1}{2} \sum_{\substack{m, n \in \mathbb{Z} \\ (m, n) \neq (0, 0)}} \frac{1}{(m\tau + n)^k |m\tau + n|^\epsilon}$ after which it is possible to define $G_2 = \lim_{\epsilon \to 0} G_{2,\epsilon}$. Then all G_k, $k \in 2\mathbb{Z}$, $k \geq 2$ have a Fourier expansion[1] in $q = \exp(2\pi i \tau)$

$$G_k(\tau) = \frac{(2\pi i)^k}{(k-1)!} \left(-\frac{B_k}{2k} + \sum_{n=1}^\infty \sigma_{k-1}(n) q^n \right), \tag{B.1.5}$$

with $\sigma_k(n) = \sum_{p|n} p^k$ the sum of kth powers of positive divisors of n and $\sum_{k=0}^\infty \frac{B_k x^k}{k!} = \frac{x}{e^x - 1}$ defining the Bernoulli numbers B_k, e.g. $B_2 = \frac{1}{6}$, $B_4 = -\frac{1}{30}$, $B_6 = \frac{1}{42}$, $B_8 = -\frac{1}{30}$, $B_{10} = \frac{5}{66}$, $B_{12} = -\frac{691}{2730}$, $B_{14} = \frac{7}{6}$ etc.

---

[1] Note that the Eisenstein series start with coefficient 1.

B.1. PSL(2,ℤ) MODULAR FORMS

Very much like in QFT the regularization introduces an anomaly in the symmetry transformation so that E_2 transforms

$$E_2(\tau_\gamma) = (c\tau + d)^2 E_2(\tau) - \frac{6ic}{\pi}(c\tau + d) \tag{B.1.6}$$

with an inhomogeneous term.

At least (E_2, E_4, E_6) form a ring, the ring of quasi modular holomorphic forms $\mathcal{M}^!$, which closes under differentiation, i.e.

$$d_\tau E_2 = \frac{1}{12}(E_2^2 - E_4), \quad d_\tau E_4 = \frac{1}{3}(E_2 E_4 - E_6), \quad d_\tau E_6 = \frac{1}{2}(E_2 E_6 - E_4^2) \,. \tag{B.1.7}$$

Using (B.1.2) and (B.1.6) we see that the inhomogeneous terms in (B.1.2,B.1.6) cancel so that

$$\hat{E}_2(\tau) = E_2(\tau) - \frac{3}{\pi \mathrm{Im}(\tau)} \tag{B.1.8}$$

transforms like a modular form of weight 2, albeit not a holomorphic one. $(\hat{E}_2, E_4, E_6)$ form the ring of almost holomorphic modular forms of Γ_1. The latter closes under the Maass derivative, which acts on forms of weight k by

$$D_\tau f_k = \left(d_\tau - \frac{k}{4\pi \mathrm{Im}(\tau)}\right) f_k \tag{B.1.9}$$

and maps $D_\tau : \mathcal{M}_k^! \to \mathcal{M}_{k+2}^!$. Note that the equations (B.1.7) hold with d_τ replaced by D_τ and $E_2(\tau)$ replaced by $\hat{E}_2(\tau)$. This Maass derivative corresponds to the covariant derivative that appears in topological string theory (3.5.35).

From the physical point of view there seems the following story behind these well known mathematical facts. The holomorphic propagator, which in the case of local geometries can be made proportional to E_2 needs some regularization, which breaks T duality. The latter is restored by adding the non-holomorphic term (B.1.8). The modular anomaly and the holomorphic anomaly are in this sense counterparts, which cannot both be realized at least perturbatively. T-duality is physically better motivated. Attempts in the literature, e.g. in an interesting paper [81], to define a holomorphic and modular non-perturbative completion by summing over orbits seem to make sense only if absolute convergence in the moduli is established, which is hard.

F^1 is an index, which is finite for smooth compact spaces. It diverges therefore only from singular configurations, that occur if e.g. the discriminant of the elliptic curve given below for the Weierstrass form $y^2 = 4x^3 - 3xE_4 + E_6$

$$\Delta(\tau) = \eta^{24}(\tau) = q \prod_{n=1}^{\infty}(1-q^n)^{24} = \frac{1}{1728}(E_4^3(\tau) - E_6^2(\tau)) \,, \tag{B.1.10}$$

vanishes. Note that the j for this curve is

$$j = 1728 \frac{E_4^3}{E_4^3 - E_6^2} = \frac{1}{q} + 744 + 196884 q + 21493760 q^2 + \mathcal{O}(q^3) \,. \tag{B.1.11}$$

160   CHAPTER B. MODULAR ANOMALY VERSUS HOLOMORPHIC ANOMALY

It follows from (B.1.3) that $\eta(\tau_\gamma) = (c\tau+d)^{\frac{1}{2}}\eta(\tau)$ transforms with weight $\frac{1}{2}$ and from (B.1.7) that

$$d_\tau \log(\eta(\tau)) = \frac{1}{24}E_2(\tau). \qquad (\text{B.1.12})$$

Further from (B.1.2) we see that $\sqrt{\text{Im}(\tau)}|\eta(\tau)|^2$ is an almost holomorphic modular invariant and from (B.1.7,B.1.8,B.1.10) that

$$d_\tau \log(\sqrt{\text{Im}(\tau)}|\eta(\tau)|^2) = \frac{1}{24}\hat{E}_2(\tau). \qquad (\text{B.1.13})$$

We need also the theta functions of general characteristic

$$\theta\begin{bmatrix}a\\b\end{bmatrix}(z,\tau) = \sum_{n\in\mathbb{Z}} \exp\left(\pi i(n+a)\tau(n+a) + 2\pi i \sum_i (z+b)n\right). \qquad (\text{B.1.14})$$

Appendix C

Details: Grassmannian Calabi-Yau manifolds

C.1 Chern classes and topological invariants

$\mathbb{G}(2,5)$: $\int_{G(2,5)} \sigma_1^6 = 5$, $\int_{G(2,5)} \sigma_2 \sigma_1^4 = 3$, $\int_{G(2,5)} \sigma_3 \sigma_1^3 = 1$,

$(\mathbb{G}(2,5)\|1,1,3)^1_{-150}$: $c((\mathbb{G}(2,5)\|1,1,3)^1_{-150})$
$= 1 + (5c_1(Q)^2 - c_2(Q))$
$- (8c_1(Q)^3 + 5c_1(Q)c_2(Q) - 5c_3(Q)) + \cdots$,

$\Rightarrow \chi = -150$, $c_2 \cdot H = 66$, $H^3 = 15$.

$(\mathbb{G}(2,5)\|1,2,2)^1_{-120}$: $c((\mathbb{G}(2,5)\|1,2,2)^1_{-120})$
$= 1 + (4c_1(Q)^2 - c_2(Q))$
$- (4c_1(Q)^3 + 5c_1(Q)c_2(Q) - 5c_3(Q)) + \cdots$,

$\Rightarrow \chi = -120$, $c_2 \cdot H = 68$, $H^3 = 20$.

$\mathbb{G}(2,6)$: $\int_{G(2,6)} \sigma_1^8 = 14$, $\int_{G(2,6)} \sigma_2 \sigma_1^6 = 9$, $\int_{G(2,6)} \sigma_3 \sigma_1^5 = 4$,

$(\mathbb{G}(2,6)\|1,1,1,1,2)^1_{-116}$: $c((\mathbb{G}(2,6)\|1,1,1,1,2)^1_{-116})$
$= 1 + (4c_1(Q)^2 - 2c_2(Q))$
$- (2c_1(Q)^3 + 6c_1(Q)c_2(Q) - 6c_3(Q)) + \cdots$,

$\Rightarrow \chi = -116$, $c_2 \cdot H = 76$, $H^3 = 28$.

$\mathbb{G}(3,6)$: $\int_{G(3,6)} \sigma_1^9 = 42$, $\int_{G(3,6)} \sigma_2 \sigma_1^7 = 21$, $\int_{G(3,6)} \sigma_3 \sigma_1^6 = 5$,

$(\mathbb{G}(3,6)\|1^6)^1_{-96}$: $c((\mathbb{G}(3,6)\|1^6)^1_{-96})$
$= 1 + 2c_1(Q)^2$
$- (6c_1(Q)c_2(Q) - 6c_3(Q)) + \cdots$,

$\Rightarrow \chi = -96$, $c_2 \cdot H = 84$, $H^3 = 42$.

$\mathbb{G}(2,7)$: $\int_{G(2,7)} \sigma_1^{10} = 42$, $\int_{G(2,7)} \sigma_2 \sigma_1^8 = 28$, $\int_{G(2,7)} \sigma_3 \sigma_1^7 = 14$,

$(\mathbb{G}(2,7)\|1^7)^1_{-98}$: $c((\mathbb{G}(2,7)\|1^7)^1_{-98})$
$= 1 + (4c_1(Q)^2 - 3c_2(Q))$
$- (7c_1(Q)c_2(Q) - 7c_3(Q)) + \cdots$,

$\Rightarrow \chi = -98$, $c_2 \cdot H = 84$, $H^3 = 42$.

C.2 Tables of Gopakumar-Vafa invariants

d	$g=0$	$g=1$	$g=2$
1	540	0	0
2	12555	0	0
3	621315	-1	0
4	44892765	13095	0
5	3995437590	17230617	-1080
6	406684089360	6648808835	921735
7	45426958360155	1831575868830	6512362740
8	5432556927598425	433375127634753	5837267557035
9	684486974574277695	94416986839804040	3061620003073095
10	89872619976165978675	19571240651198871015	1223886411726167880

d	$g=3$	$g=4$	$g=5$
1	0	0	0
2	0	0	0
3	0	0	0
4	0	0	0
5	0	0	0
6	420	5	0
7	-26460	-2160	0
8	6528493485	218160	-2160
9	20216637579465	6735865790	2770635
10	22818718255545315	85314971897190	5441786955

Table C.2.1: Gopakumar-Vafa invariants $n_g(d)(g \leq 5)$ of the Grassmannian Calabi-Yau threefold $(\mathbb{G}(2,5)\|1,1,3)^1_{-150}$.

d	$g=0$	$g=1$	$g=2$	$g=3$	$g=4$	$g=5$
1	400	0	0	0	0	0
2	5540	0	0	0	0	0
3	164400	0	0	0	0	0
4	7059880	1537	0	0	0	0
5	373030720	882496	0	0	0	0
6	22532253740	214941640	15140	0	0	0
7	1493352046000	37001766880	57840400	-800	0	0
8	105953648564840	5388182343297	36620960080	10792630	320	5
9	7919320425000000	715201587952800	12817600017680	33952864320	697600	-1600
10	616905540794800	89732472170109248	3295358054573602	2938605942400	32052405340	-32320

Table C.2.2: Gopakumar-Vafa invariants $n_g(d) (g \leq 5)$ of the Grassmannian Calabi-Yau threefold $(\mathbb{G}(2,5) \| 1, 2, 2)^1_{-120}$.

C.2. TABLES OF GOPAKUMAR-VAFA INVARIANTS

d	$g=0$	$g=1$	$g=2$	$g=3$	$g=4$	$g=5$
1	210	0	0	0	0	0
2	1176	0	0	0	0	0
3	13104	0	0	0	0	0
4	201936	0	0	0	0	0
5	3824016	84	0	0	0	0
6	82568136	74382	0	0	0	0
7	1954684008	8161452	0	0	0	0
8	49516091520	560512344	70896	0	0	0
9	1321186053432	31354814820	39198978	0	0	0
10	36729091812168	1568818990200	7239273552	1086246	0	0
11	1055613263065704	73339159104540	827701960638	932836632	1722	0
12	31184875579315920	3279169536538154	72679697259288	284870410986	55653752	0

Table C.2.3: Gopakumar-Vafa invariants $n_g(d)$ ($g \leq 5$) of the Grassmannian Calabi-Yau threefold $(\mathbb{G}(3,6)\|1^6)^{-1}_{-96}$.

d	$g=0$	$g=1$	$g=2$	$g=3$	$g=4$	$g=5$
1	280	0	0	0	0	0
2	2674	0	0	0	0	0
3	48272	0	0	0	0	0
4	1279040	27	0	0	0	0
5	41389992	26208	0	0	0	0
6	1531603276	5914124	-54	0	0	0
7	62153423432	745052912	56112	0	0	0
8	2699769672096	73219520613	120462612	-5267	0	0
9	123536738915800	6326648922384	40927354944	4713072	840	0
10	5890247824324990	506932941439940	8145450103430	15699104736	-91464	-404
11	290364442225572848	38717395881042032	1228133118935408	8307363701728	4174512664	66640
12	14713407331980050400	2863231551878100494	156147718274297768	2460694451990694	75347873089 68	991403118

Table C.2.4: Gopakumar-Vafa invariants $n_g(d)(g \leq 5)$ of the Grassmannian Calabi-Yau threefold $(\mathbb{G}(2,6)\|1,1,1,2)^1_{-116}$.

C.2. TABLES OF GOPAKUMAR-VAFA INVARIANTS

d	$g=0$	$g=1$	$g=2$	$g=3$	$g=4$	$g=5$
1	196	0	0	0	0	0
2	1225	0	0	0	0	0
3	12740	0	0	0	0	0
4	198058	0	0	0	0	0
5	3716944	588	0	0	0	0
6	79823205	99960	0	0	0	0
7	1877972628	8964372	0	0	0	0
8	47288943912	577298253	99960	0	0	0
9	1254186001124	31299964612	47151720	-1176	0	0
10	34657942457488	1535808070650	7906245550	325409	0	0
11	990133717028596	70785403788680	858740761340	956485684	-25480	3675
12	29075817464070412	3129139504135680	73056658523632	30122732311	27885116	73892

Table C.2.5: Gopakumar-Vafa invariants $n_g(d) (g \leq 5)$ of the Grassmannian Calabi-Yau threefold $(\mathbb{G}(2,7)\|1^7)^1_{-98}$.

d	$g=0$	$g=1$	$g=2$	$g=3$
1	588	0	0	0
2	12103	0	0	0
3	583884	196	0	0
4	41359136	99960	0	0
5	3609394096	34149668	12740	0
6	360339083307	9220666238	25275866	1225
7	39487258327356	2163937552736	21087112172	22409856
8	4633258198646014	466455116030169	11246111235996	58503447590
9	572819822939575596	9535308920590736	46010048597770928	6777902782204
10	73802503401477453288	188297534581341112872	15867773907506411117	500692818827807 27

d	$g=4$	$g=5$
1	0	0
2	0	0
3	0	0
4	0	0
5	0	0
6	0	0
7	0	0
8	25371416	3675
9	216888021056	33575388
10	521484626374894	111178828 6385

Table C.2.6: Gopakumar-Vafa invariants $n_g(d)(g \leq 5)$ of the Pfaffian Calabi-Yau threefold M'.

C.3 5D black hole asymptotic

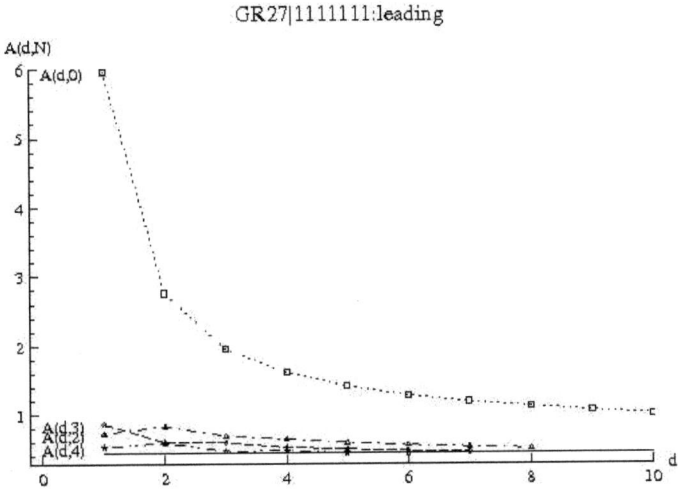

Figure C.1: Leading behavior of the microscopic entropy for the 5d black hole for the Grassmannian Calabi-Yau threefold $(\mathbb{G}(2,7)\|1,1,1,1,1,1,1)^1_{-98}$. $A(d,m)$ are the Richardson transforms. The Richardson transforms of the microscopic entropy converge within 4 % to the expected value from the macroscopic calculation $b_0 = \frac{4\pi}{3\sqrt{2H^3}} \sim .046$ for $H^3 = 42$, see [66] for details.

Appendix D
Details: Local Calabi-Yau manifolds

D.1 Gopakumar-Vafa invariants of local Calabi-Yau manifolds

d_1	0	1	2	3	4	5	6
d_2							
0		-2	0	0	0	0	0
1	-2	-4	-6	-8	-10	-12	-14
2	0	-6	-32	-110	-288	-644	-1280
3	0	-8	-110	-756	-3556	-13072	-40338
4	0	-10	-288	-3556	-27264	-153324	-690400
5	0	-12	-644	-13072	-153324	-1252040	-7877210
6	0	-14	-1280	-40338	-690400	-7877210	-67008672

Table D.1.1: Instanton numbers $n_{d_1 d_2}^{g=0}$ of local $\mathbb{K}_{\mathbb{F}_0}$

d_1	0	1	2	3	4	5	6
d_2							
0		0	0	0	0	0	0
1	0	0	0	0	0	0	0
2	0	0	9	68	300	988	2698
3	0	0	68	1016	7792	41376	172124
4	0	0	300	7792	95313	760764	4552692
5	0	0	988	41376	760764	8695048	71859628
6	0	0	2698	172124	4552692	71859628	795165949

Table D.1.2: Genus one GV invariants $n_{d_1 d_2}^{g=1}$ of local $\mathbb{K}_{\mathbb{F}_0}$

d_1	0	1	2	3	4	5	6
d_2							
0		0	0	0	0	0	0
1	0	0	0	0	0	0	0
2	0	0	0	-12	-116	-628	-2488
3	0	0	-12	-580	-8042	-64624	-371980
4	0	0	-116	-8042	-167936	-1964440	-15913228
5	0	0	-628	-64624	-1964440	-32242268	-355307838
6	0	0	-2488	-371980	-15913228	-355307838	-5182075136

Table D.1.3: Genus two GV invariants $n_{d_1 d_2}^{g=2}$ of local $\mathbb{K}_{\mathbb{F}_0}$

d_1	0	1	2	3	4	5	6
d_2							
0		0	0	0	0	0	0
1	0	0	0	0	0	0	0
2	0	0	0	0	15	176	1130
3	0	0	0	156	4680	60840	501440
4	0	0	15	4680	184056	3288688	36882969
5	0	0	176	60840	3288688	80072160	1198255524
6	0	0	1130	501440	36882969	1198255524	23409326968

Table D.1.4: Genus three GV invariants $n_{d_1 d_2}^{g=3}$ of local $\mathbb{K}_{\mathbb{F}_0}$

d_1	0	1	2	3	4	5	6
d_2							
0		0	0	0	0	0	0
1	0	0	0	0	0	0	0
2	0	0	0	0	0	-18	-248
3	0	0	0	-16	-1560	-36408	-450438
4	0	0	0	-1560	-133464	-3839632	-61250176
5	0	0	-18	-36408	-3839632	-144085372	-2989287812
6	0	0	-248	-450438	-61250176	-2989287812	-79635105296

Table D.1.5: Genus four GV invariants $n_{d_1 d_2}^{g=4}$ of local $\mathbb{K}_{\mathbb{F}_0}$

d_1	0	1	2	3	4	5	6	7
d_2								
0		-2	0	0	0	0	0	0
1	1	3	5	7	9	11	13	15
2	0	0	-6	-32	-110	-288	-644	-1280
3	0	0	0	27	286	1651	6885	23188
4	0	0	0	0	-192	-3038	-25216	-146718
5	0	0	0	0	0	1695	35870	392084
6	0	0	0	0	0	0	-17064	-454880
7	0	0	0	0	0	0	0	188454

Table D.1.6: Instanton numbers $n_{d_1 d_2}^{g=0}$ of local $\mathbb{K}_{\mathbb{F}_1}$

d_1	0	1	2	3	4	5	6	7
d_2								
0		0	0	0	0	0	0	0
1	0	0	0	0	0	0	0	0
2	0	0	0	9	68	300	988	2698
3	0	0	0	-10	-288	-2938	-18470	-86156
4	0	0	0	0	231	6984	90131	736788
5	0	0	0	0	0	-4452	-152622	-2388864
6	0	0	0	0	0	0	80948	3164814
7	0	0	0	0	0	0	0	-1438086

Table D.1.7: Genus one GV invariants $n_{d_1 d_2}^{g=1}$ of local $\mathbb{K}_{\mathbb{F}_1}$

d_1	0	1	2	3	4	5	6	7
d_2								
0		0	0	0	0	0	0	0
1	0	0	0	0	0	0	0	0
2	0	0	0	0	-12	-116	-628	-2488
3	0	0	0	0	108	2353	23910	160055
4	0	0	0	0	-102	-7506	-161760	-1921520
5	0	0	0	0	0	5430	329544	7667739
6	0	0	0	0	0	0	-194022	-11643066
7	0	0	0	0	0	0	0	5784837

Table D.1.8: Genus two GV invariants $n_{d_1 d_2}^{g=2}$ of local $\mathbb{K}_{\mathbb{F}_1}$

d_1	0	1	2	3	4	5	6	7
d_2								
0		0	0	0	0	0	0	0
1	0	0	0	0	0	0	0	0
2	0	0	0	0	0	15	176	1130
3	0	0	0	0	-14	-992	-18118	-182546
4	0	0	0	0	15	4519	179995	3243067
5	0	0	0	0	0	-3672	-447502	-16230032
6	0	0	0	0	0	0	290853	28382022
7	0	0	0	0	0	0	0	-15363990

Table D.1.9: Genus three GV invariants $n_{d_1 d_2}^{g=3}$ of local $\mathbb{K}_{\mathbb{F}_1}$

Appendix E

Details: K3 Fibrations

E.1 Gopakumar-Vafa invariants

d_1	0	1	2	3	4	5	6
d_2							
0	0	640	10032	288384	10979984	495269504	24945542832
1	4	640	72224	7539200	757561520	74132328704	7117563990784
2	0	0	10032	7539200	2346819520	520834042880	95728361673920
3	0	0	0	288384	757561520	520834042880	212132862927264
4	0	0	0	0	10979984	74132328704	95728361673920
5	0	0	0	0	0	495269504	7117563990784
6	0	0	0	0	0	0	24945542832

Table E.1.1: Instanton numbers $n_{d_1 d_2}^{g=0}$ of $\mathbb{P}_4^{(1,1,2,2,2)}$ [8]

d_1	0	1	2	3	4	5	6
d_2							
0	0	0	0	-1280	-317864	-36571904	-3478899872
1	0	0	0	2560	1047280	224877056	36389051520
2	0	0	0	2560	15948240	12229001216	4954131766464
3	0	0	0	-1280	1047280	12229001216	13714937870784
4	0	0	0	0	-317864	224877056	4954131766464
5	0	0	0	0	0	-36571904	36389051520
6	0	0	0	0	0	0	-3478899872

Table E.1.2: Instanton numbers $n_{d_1 d_2}^{g=1}$ of $\mathbb{P}_4^{(1,1,2,2,2)}$ [8]

d_1	0	1	2	3	4	5	6
d_2							
0	0	0	0	0	472	875392	220466160
1	0	0	0	0	-1232	-2540032	-1005368448
2	0	0	0	0	848	9699584	21816516384
3	0	0	0	0	-1232	9699584	132874256992
4	0	0	0	0	472	-2540032	21816516384
5	0	0	0	0	0	875392	-1005368448
6	0	0	0	0	0	0	220466160

Table E.1.3: Instanton numbers $n_{d_1 d_2}^{g=2}$ of $\mathbb{P}_4^{(1,1,2,2,2)}$ [8]

d_1	0	1	2	3	4	5	6
d_2							
0	0	0	0	0	8	-2560	-6385824
1	0	0	0	0	-24	3840	20133504
2	0	0	0	0	24	2560	-19124704
3	0	0	0	0	-24	2560	23433600
4	0	0	0	0	8	3840	-19124704
5	0	0	0	0	0	-2560	20133504
6	0	0	0	0	0	0	-6385824

Table E.1.4: Instanton numbers $n_{d_1 d_2}^{g=3}$ of $\mathbb{P}_4^{(1,1,2,2,2)}$ [8]

d_1	0	1	2	3	4	5	6	7
d_2								
0	0	0	0	0	0	0	50160	101090432
1	0	0	0	0	0	0	-160512	-355794944
2	0	0	0	0	0	0	220704	478526720
3	0	0	0	0	0	0	56160	-366614784
4	0	0	0	0	0	0	220704	-366614784
5	0	0	0	0	0	0	-160512	478526720
6	0	0	0	0	0	0	50160	-

Table E.1.5: Instanton numbers $n_{d_1 d_2}^{g=4}$ of $\mathbb{P}_4^{(1,1,2,2,2)}$ [8]

d_1	0	1	2	3	4	5
d_2						
0	0	2496	223752	38637504	9100224984	2557481027520
1	2	2496	1941264	1327392512	861202986072	540194037151104
2	0	0	223752	1327392512	2859010142112	4247105405354496
3	0	0	0	38637504	861202986072	4247105405354496
4	0	0	0	0	9100224984	540194037151104
5	0	0	0	0	0	2557481027520
6	0	0	0	0	0	0

Table E.1.6: Instanton numbers $n_{d_1 d_2}^{g=0}$ of $\mathbb{P}_4^{(1,1,2,2,6)}$ [12]

d_1	0	1	2	3	4	5
d_2						
0	0	0	-492	-1465984	-1042943520	-595277880960
1	0	0	480	2080000	3453856440	3900245149440
2	0	0	-492	2080000	74453838960	313232037949440
3	0	0	0	-1465984	3453856440	313232037949440
4	0	0	0	0	-1042943520	3900245149440
5	0	0	0	0	0	-595277880960

Table E.1.7: Instanton numbers $n_{d_1 d_2}^{g=1}$ of $\mathbb{P}_4^{(1,1,2,2,6)}$ [12]

d_1	0	1	2	3	4	5
d_2						
0	0	0	-6	7488	50181180	72485905344
1	0	0	8	0	-73048296	-194629721856
2	0	0	-6	0	32635544	2083061531520
3	0	0	0	7488	-73048296	2083061531520
4	0	0	0	0	50181180	-194629721856
5	0	0	0	0	0	72485905344

Table E.1.8: Instanton numbers $n_{d_1 d_2}^{g=2}$ of $\mathbb{P}_4^{(1,1,2,2,6)}$ [12]

d_1	0	1	2	3	4	5
d_2						
0	0	0	0	0	-902328	-5359699200
1	0	0	0	0	1357500	10139497472
2	0	0	0	0	-822968	7645673856
3	0	0	0	0	1357500	-7645673856
4	0	0	0	0	-902328	10139497472
5	0	0	0	0	0	-5359699200

Table E.1.9: Instanton numbers $n_{d_1 d_2}^{g=3}$ of $\mathbb{P}_4^{(1,1,2,2,6)}$ [12]

d_1	0	1	2	3	4	5
d_2						
0	0	0	0	0	1164	228623232
1	0	0	0	0	-1820	-376523648
2	0	0	0	0	2768	144351104
3	0	0	0	0	-1820	144351104
4	0	0	0	0	1164	-376523648
5	0	0	0	0	0	228623232

Table E.1.10: Instanton numbers $n_{d_1 d_2}^{g=4}$ of $\mathbb{P}_4^{(1,1,2,2,6)}$[12]

Bibliography

[1] B. R. Greene, D. R. Morrison and A. Strominger, "Black hole condensation and the unification of string vacua," Nucl. Phys. B **451**, 109 (1995) [arXiv:hep-th/9504145].

[2] H. Ooguri, A. Strominger and C. Vafa, Phys. Rev. D **70**, 106007 (2004) [arXiv:hep-th/0405146].

[3] B. Haghighat and A. Klemm, "Topological Strings on Grassmannian Calabi-Yau manifolds," arXiv:0802.2908 [hep-th].

[4] B. Haghighat, A. Klemm and M. Rauch, "Integrability of the holomorphic anomaly equations," JHEP **0810**, 097 (2008) [arXiv:0809.1674 [hep-th]].

[5] B. Haghighat and A. Klemm, arXiv:0908.0336 [hep-th].

[6] J. Polchinski, "String theory. Vol. 1: An introduction to the bosonic string," *Cambridge, UK: Univ. Pr. (1998) 402 p*

[7] J. Polchinski, "String theory. Vol. 2: An introduction to the bosonic string," *Cambridge, UK: Univ. Pr. (1998) 402 p*

[8] P. S. Aspinwall, "K3 surfaces and string duality," arXiv:hep-th/9611137.

[9] M. Bodner, A. C. Cadavid and S. Ferrara, "(2,2) vacuum configurations for type IIA superstrings: N=2 supergravity Lagrangians and algebraic geometry," Class. Quant. Grav. **8**, 789 (1991).

[10] J. D. Lykken, "Introduction to supersymmetry," arXiv:hep-th/9612114.

[11] N. Seiberg and E. Witten, "Monopole Condensation, And Confinement In N=2 Supersymmetric Yang-Mills Theory," Nucl. Phys. B **426**, 19 (1994) [Erratum-ibid. B **430**, 485 (1994)] [arXiv:hep-th/9407087].

[12] N. Seiberg, "Supersymmetry And Nonperturbative Beta Functions," Phys. Lett. B **206**, 75 (1988).

[13] S. H. Katz, A. Klemm and C. Vafa, "Geometric engineering of quantum field theories," Nucl. Phys. B **497**, 173 (1997) [arXiv:hep-th/9609239].

[14] E. Witten, "String theory dynamics in various dimensions," Nucl. Phys. B **443**, 85 (1995) [arXiv:hep-th/9503124].

[15] A. Strominger and C. Vafa, Phys. Lett. B **379**, 99 (1996) [arXiv:hep-th/9601029].

[16] S. Ferrara, R. Kallosh and A. Strominger, "N=2 extremal black holes," Phys. Rev. D **52**, 5412 (1995) [arXiv:hep-th/9508072].

[17] . Kallosh, A. Rajaraman and W. K. Wong, Phys. Rev. D **55**, 3246 (1997) [arXiv:hep-th/9611094].

[18] R. M. Wald, "Black hole entropy is the Noether charge," Phys. Rev. D **48**, 3427 (1993) [arXiv:gr-qc/9307038].

[19] S. H. Katz, A. Klemm and C. Vafa, "M-theory, topological strings and spinning black holes," Adv. Theor. Math. Phys. **3**, 1445 (1999) [arXiv:hep-th/9910181].

[20] A. Strominger, Nucl. Phys. B **451**, 96 (1995) [arXiv:hep-th/9504090].

[21] P. Griffiths and J. Harris, "Principles of Algebraic Geometry".

[22] D. Huybrechts, "Complex Geometry", Springer Verlag

[23] A. Strominger, "Yukawa Couplings In Superstring Compactification," Phys. Rev. Lett. **55**, 2547 (1985).

[24] P. Candelas and X. de la Ossa, "MODULI SPACE OF CALABI-YAU MANIFOLDS," Nucl. Phys. B **355**, 455 (1991).

[25] R. Bryant and P. Griffiths, Progress in Mathematics **36** pp. 77-102, (Birkhäuser, Boston, 1983).

[26] B. R. Greene, "String theory on Calabi-Yau manifolds," arXiv:hep-th/9702155.

[27] W. Lerche, C. Vafa and N. P. Warner, "ADDENDUM TO 'CHIRAL RINGS IN N=2 SUPERCONFORMAL THEORIES',"

[28] P. H. Ginsparg, "APPLIED CONFORMAL FIELD THEORY," arXiv:hep-th/9108028.

[29] B. Zumino, "Supersymmetry And Kahler Manifolds," Phys. Lett. B **87**, 203 (1979).

[30] L. J. Dixon, "SOME WORLD SHEET PROPERTIES OF SUPERSTRING COMPACTIFICATIONS, ON ORBIFOLDS AND OTHERWISE,"

[31] J. Distler and B. R. Greene, "Some Exact Results on the Superpotential from Calabi-Yau Compactifications," Nucl. Phys. B **309**, 295 (1988).

[32] E. Witten, "Phases of N = 2 theories in two dimensions," Nucl. Phys. B **403**, 159 (1993) [arXiv:hep-th/9301042].

[33] P. S. Aspinwall, B. R. Greene and D. R. Morrison, "Calabi-Yau moduli space, mirror manifolds and spacetime topology change in string theory," Nucl. Phys. B **416**, 414 (1994) [arXiv:hep-th/9309097].

[34] E. Witten, "Supersymmetry and Morse theory," J. Diff. Geom. **17** (1982) 661.

[35] E. Witten, "Mirror manifolds and topological field theory," arXiv:hep-th/9112056.

[36] A. Klemm, "Topological String Theory On Calabi-Yau Threefolds," PoS **RTN2005** (2005) 002.

[37] L. Alvarez-Gaume, C. Gomez, G. W. Moore and C. Vafa, "Strings In The Operator Formalism," Nucl. Phys. B **303**, 455 (1988).

[38] R. Dijkgraaf, "Intersection theory, integrable hierarchies and topological field theory," arXiv:hep-th/9201003.

[39] P. S. Aspinwall, B. R. Greene and D. R. Morrison, "Space-time topology change and stringy geometry," J. Math. Phys. **35**, 5321 (1994).

[40] M. Bershadsky, S. Cecotti, H. Ooguri and C. Vafa, "Kodaira-Spencer theory of gravity and exact results for quantum string Commun. Math. Phys. **165**, 311 (1994) [arXiv:hep-th/9309140].

[41] I. Antoniadis, E. Gava, K. S. Narain and T. R. Taylor, "Topological amplitudes in string theory," Nucl. Phys. B **413**, 162 (1994) [arXiv:hep-th/9307158].

[42] R. Gopakumar and C. Vafa, "M-theory and topological strings. I," arXiv:hep-th/9809187.

[43] R. Gopakumar and C. Vafa, "M-theory and topological strings. II," arXiv:hep-th/9812127.

[44] C. Itzykson, J. Zuber, "Quantum field theory", Addison Wesley Publishing

[45] K. Hori et al., "Mirror symmetry," *Providence, USA: AMS (2003) 929 p*

[46] M. Bershadsky, S. Cecotti, H. Ooguri and C. Vafa, "Holomorphic anomalies in topological field theories," Nucl. Phys. B **405**, 279 (1993) [arXiv:hep-th/9302103].

[47] S. Yamaguchi and S. T. Yau, "Topological string partition functions as polynomials," JHEP **0407**, 047 (2004) [arXiv:hep-th/0406078].

[48] T. W. Grimm, A. Klemm, M. Marino and M. Weiss, "Direct integration of the topological string," JHEP **0708**, 058 (2007) [arXiv:hep-th/0702187].

[49] M. Alim and J. D. Lange, "Polynomial Structure of the (Open) Topological String Partition Function," JHEP **0710**, 045 (2007) [arXiv:0708.2886 [hep-th]].

[50] T. W. Grimm, A. Klemm, M. Marino and M. Weiss, JHEP **0708**, 058 (2007) [arXiv:hep-th/0702187].

[51] M. x. Huang, A. Klemm and S. Quackenbush, "Topological String Theory on Compact Calabi-Yau: Modularity and Boundary Conditions," arXiv:hep-th/0612125.

[52] D. Zagier, " Elliptic modular forms and their applications," University Text, Springer, Heidelberg (2007).

[53] R. Gopakumar and C. Vafa, "Branes and Fundamental Groups", [arXiv:hep-th/9712048].

[54] C. Vafa, "A Stringy test of the fate of the conifold," Nucl. Phys. B **447**, 252 (1995) [arXiv:hep-th/9505023].

[55] G. Tian, "Smoothness of the Universal Deformaton Space of Compact Calabi-Yau manifolds and its Weil-Peterson metric," Mathematical Aspects of String Theory, Ed. S.T. Yau, World Scientific Singapore (1987).

[56] A. Todorov, "The Weyl-Peterson Geometry of the Moduli-Space of $SU(n \geq 3)$ (Calabi-Yau) Manifolds I," Commun. Math. Phys. 126 (1989) 325.

[57] D. R. Morrison, "Picard-Fuchs equations and mirror maps for hypersurfaces," arXiv:hep-th/9111025.

[58] D. A. Cox and S. Katz, "Mirror symmetry and algebraic geometry," *Providence, USA: AMS (2000) 469 p*

[59] V. Batyrev, I. Ciocan-Fontanine, B. Kim and D. v. Straten, "Conifold Transitions and Mirror Symmetry for Calabi-Yau Complete Intersections in Grassmannians," alg-geom/9710022.

[60] A. Borel and F. Hirzebruch, "Characteristic classes and homogeneous spaces I", Amer. J. Math. 80 1958 458-538.

[61] B. Sturmfels, "Gröbner Bases and Convex Polytopes", Univ. Lect. Notes, vo. 8, AMS, 1996.

[62] E. L. Ince, "Ordinary differential equations," Dover Publications (June 1, 1956).

[63] S. Hosono and Y. Konishi, "Higher genus Gromov-Witten invariants of the Grassmannian, and the Pfaffian Calabi-Yau threefolds", [arXiv:math.AG/0704.2928].

[64] R. Gopakumar and C. Vafa, "Branes and Fundamental Groups", [arXiv:hep-th/9712048].

[65] C. van Enckevort and D. van Straten, "Monodromy calculations of fourth order equations of Calabi-Yau type", [arXiv:math.AG/0412539].

[66] M. x. Huang, A. Klemm, M. Marino and A. Tavanfar, "Black Holes and Large Order Quantum Geometry," arXiv:0704.2440 [hep-th].

[67] E.A. Rodland, "The Pfaffian Calabi-Yau, its Mirror and their link to the Grassmannian $G(2,7)$", Compositio Math. 122 (2000), no. 2, 135 - 149, [arXiv:math.AG/9801092].

[68] J. Bryan, R. Pandharipande, "Curves in Calabi-Yau 3-folds and Topological Quantum Field Theory," arXiv:math/0306316

[69] A. Klemm, P. Mayr and C. Vafa, "BPS states of exceptional non-critical strings," arXiv:hep-th/9607139.

[70] T. M. Chiang, A. Klemm, S. T. Yau and E. Zaslow, "Local mirror symmetry: Calculations and interpretations," Adv. Theor. Math. Phys. **3**, 495 (1999) [arXiv:hep-th/9903053].

[71] K. Hori and C. Vafa, "Mirror symmetry," arXiv:hep-th/0002222.

[72] K. Hori et al., "Mirror symmetry," *Providence, USA: AMS (2003) 929 p*

[73] W. Fulton, "Introduction to toric varieties", Princeton Univ. Press, Princeton (1993).

[74] D.A. Cox, "The Homogeneous Coordinate Ring of a Toric Variety," J. Algebraic Geom. 4 17 (1995) [arXiv:alg-geom/9210008].

[75] P. S. Aspinwall, B. R. Greene and D. R. Morrison, "Space-time topology change and stringy geometry," J. Math. Phys. **35** (1994) 5321.

[76] A. Klemm and E. Zaslow, "Local mirror symmetry at higher genus," arXiv:hep-th/9906046.

[77] M. Aganagic, A. Klemm, M. Marino and C. Vafa, "Matrix model as a mirror of Chern-Simons theory," JHEP **0402** (2004) 010 [arXiv:hep-th/0211098].

[78] S. Hosono, A. Klemm, S. Theisen and S. T. Yau, "Mirror Symmetry, Mirror Map And Applications To Calabi-Yau Hypersurfaces," Commun. Math. Phys. **167** (1995) 301 [arXiv:hep-th/9308122].

[79] M. Aganagic, M. Marino and C. Vafa, "All loop topological string amplitudes from Chern-Simons theory," Commun. Math. Phys. **247** (2004) 467 [arXiv:hep-th/0206164].

[80] M. Aganagic, V. Bouchard and A. Klemm, "Topological Strings and (Almost) Modular Forms," Commun. Math. Phys. **277**, 771 (2008) [arXiv:hep-th/0607100].

[81] B. Eynard, "Large N expansion of convergent matrix integrals, holomorphic anomalies, and background independence," arXiv:0802.1788 [math-ph].

[82] P. Candelas, X. De La Ossa, A. Font, S. H. Katz and D. R. Morrison, "Mirror symmetry for two parameter models. I," Nucl. Phys. B **416**, 481 (1994) [arXiv:hep-th/9308083].

[83] V. V. Batyrev, "Dual polyhedra and mirror symmetry for Calabi-Yau hypersurfaces in toric varieties," J. Alg. Geom. **3** (1994) 493.

[84] P. Griffiths, Ann. of Math. 90 (1969) 460

[85] I.M. Gel'fand, A.V. Zelevinsky and M.M. Kapranov, Func. Anal. Appl. 28 (1989) 12 and Adv. Math 84 (1990) 255

[86] S. Hosono, A. Klemm, S. Theisen and S. T. Yau, "Mirror symmetry, mirror map and applications to complete intersection Calabi-Yau spaces," Nucl. Phys. B **433**, 501 (1995) [arXiv:hep-th/9406055].

[87] S. Kachru and C. Vafa, "Exact results for N = 2 compactifications of heterotic strings," Nucl. Phys. Proc. Suppl. **46**, 210 (1996).

[88] S. Ferrara and A. Van Proeyen, "A THEOREM ON N=2 SPECIAL KAHLER PRODUCT MANIFOLDS," Class. Quant. Grav. **6**, L243 (1989).

[89] K. Oguiso, "On algebraic Fiber space structures on a Calabi-Yau 3-fold", Int. J. of Math. 4 (1993) 439

[90] P. S. Aspinwall and J. Louis, "On the Ubiquity of K3 Fibrations in String Duality," Phys. Lett. B **369**, 233 (1996) [arXiv:hep-th/9510234].

[91] A. Klemm and P. Mayr, "Strong Coupling Singularities and Non-abelian Gauge Symmetries in $N = 2$ String Theory," Nucl. Phys. B **469**, 37 (1996) [arXiv:hep-th/9601014].

[92] S. H. Katz, D. R. Morrison and M. Ronen Plesser, "Enhanced Gauge Symmetry in Type II String Theory," Nucl. Phys. B **477**, 105 (1996) [arXiv:hep-th/9601108].

[93] I. Antoniadis, E. Gava, K. S. Narain and T. R. Taylor, "Topological amplitudes in string theory," Nucl. Phys. B **413**, 162 (1994)

[94] I. Antoniadis, E. Gava, K. S. Narain and T. R. Taylor, "N=2 type II heterotic duality and higher derivative F terms," Nucl. Phys. B **455**, 109 (1995) [arXiv:hep-th/9507115].

[95] M. Marino and G. W. Moore, "Counting higher genus curves in a Calabi-Yau manifold," Nucl. Phys. B **543**, 592 (1999) [arXiv:hep-th/9808131].

[96] A. Klemm, M. Kreuzer, E. Riegler and E. Scheidegger, "Topological string amplitudes, complete intersection Calabi-Yau spaces and threshold corrections," JHEP **0505**, 023 (2005) [arXiv:hep-th/0410018].

[97] D. Maulik and R. Pandharipande, "Gromov-Witten theory and Noether-Lefshetz theory," [arXiv/0705.1653].

[98] R. Borcherds, "The Gross-Kohnen-Zagier theorem in higher dimensions." Duke J. Math **97** (1999) 219-233.

[99] T. Kawai, "String duality and modular forms", Phys. Lett. B **397** (1997) 51 [arXiv:hep-th/9607078]

[100] D. Zagier, "Traces of singular moduli," Motives, Polylogarithms and Hodge Theory, Part I, Irvine CA 1998 in Int. Press Lect. Ser. **3**(I), Int. Press. Sommerville, MA, 211-244.

[101] S. Kachru, A. Klemm, W. Lerche, P. Mayr and C. Vafa, "Nonperturbative results on the point particle limit of N=2 heterotic string compactifications," Nucl. Phys. B **459**, 537 (1996) [arXiv:hep-th/9508155].

[102] D. Gepner, "Spacetime supersymmetries in compactified string theory and superconformal models", Nucl. Phys. B **296**, 757.

[103] B. R. Greene, C. Vafa and N. P. Warner, "Calabi-Yau Manifolds and Renormalization Group Flows," Nucl. Phys. B **324**, 371 (1989).

[104] G. Curio, A. Klemm, D. Lust and S. Theisen, "On the vacuum structure of type II string compactifications on Calabi-Yau spaces with H-fluxes," Nucl. Phys. B **609**, 3 (2001) [arXiv:hep-th/0012213].

[105] A. B. Givental, "Elliptic Gromov - Witten invariants and the generalized mirror conjecture," Integrable systems and algebraic geometry (Kobe/Kyoto, 1997), 107, World Sci. Publishing, River Edge, NJ, 1998, [math.AG/9803053] .

[106] Y. Kazama and H. Suzuki, "Characterization Of N=2 Superconformal Models Generated By Coset Space Method," Phys. Lett. B **216**, 112 (1989).

[107] M. Marino, JHEP **0803**, 060 (2008) [arXiv:hep-th/0612127].

[108] V. Bouchard, A. Klemm, M. Marino and S. Pasquetti, Commun. Math. Phys. **287**, 117 (2009) [arXiv:0709.1453 [hep-th]].

I want morebooks!

Buy your books fast and straightforward online - at one of the world's fastest growing online book stores! Environmentally sound due to Print-on-Demand technologies.

Buy your books online at
www.get-morebooks.com

Kaufen Sie Ihre Bücher schnell und unkompliziert online – auf einer der am schnellsten wachsenden Buchhandelsplattformen weltweit!
Dank Print-On-Demand umwelt- und ressourcenschonend produziert.

Bücher schneller online kaufen
www.morebooks.de

OmniScriptum Marketing DEU GmbH
Heinrich-Böcking-Str. 6-8
D - 66121 Saarbrücken
Telefax: +49 681 93 81 567-9

info@omniscriptum.com
www.omniscriptum.com

Printed by Books on Demand GmbH, Norderstedt / Germany